Sustaining and Scaling Up Community Managed Water

Praise for this book

'Although the design, construction and initial capital investment of new water supply schemes pose some challenges, these are trivial compared to the subsequent issues around management, financing and sustainability of the services this infrastructure provides.

These post-construction issues are rightly receiving much attention in water sector debates and dialogues at the present time. Questions are being asked about the viability of alternative management arrangements, about the affordability of high levels of service to both water users and the State, and consequently about how sustainable water supply services can be achieved.

This is therefore an important book. It addresses these questions in a focused and in-depth manner, in relation to a single but extensive programme in a single country. The studies and investigations on which it is based are thorough and rigorous, and the arguments in the book are set out clearly for consideration by rural water specialists beyond the programme and country which are its focus.

If you are struggling with the issues just mentioned – that are covered in far greater depth in this book – I do urge you to read it, reflect on the lessons drawn by the authors, debate them, contextualise them and apply them in your work.'

Professor Richard Carter, Consultant and specialist

Sustaining and Scaling Up Community Managed Water

WASEP in Pakistan

Edited by
Jeff Tan, Stephen M. Lyon, and Attaullah Shah

Practical Action Publishing Ltd
25 Albert Street, Rugby,
Warwickshire, CV21 2SD, UK
www.practicalactionpublishing.com

A catalogue record for this book is available from the British Library.
A catalogue record for this book has been requested from the Library of Congress.

ISBN 978-1-78853-418-5 Paperback
ISBN 978-1-78853-420-8 Electronic book

Citation: Tan, J. (2025) *Sustaining and Scaling Up Community Managed Water: WASEP in Pakistan*, Rugby, UK: Practical Action Publishing http://doi.org/10.3362/9781788534208

Since 1974, Practical Action Publishing has published and disseminated books and information in support of international development work throughout the world.

Practical Action Publishing is a trading name of Practical Action Publishing Ltd (Company Reg. No. 1159018), the wholly owned publishing company of Practical Action. Practical Action Publishing trades only in support of its parent charity objectives and any profits are covenanted back to Practical Action (Charity Reg. No. 247257, Group VAT Registration No. 880 9924 76).

Cover photo shows: WASEP pipe laying in Danyore, Gilgit
Cover photo credit: Jeff Tan
Cover design by Katarzyna Markowska, Practical Action Publishing
Typeset by vPrompt eServices, India

The manufacturer's authorised representative in the EU for product safety is Lightning Source France, 1 Av. Johannes Gutenberg, 78310 Maurepas, France.
compliance@lightningsource.fr

Contents

Preface

This edited volume represents the findings from a two-year research project on the sustainability and scalability of the Water and Sanitation Extension Programme (WASEP) in Gilgit-Baltistan funded by The British Academy's Urban Infrastructures for Well-Being 2019 programme. As an edited volume it is unusual in terms of the consistent formatting and common thread that runs through each chapter. These are meant to provide the coherence of a book as opposed to a collection of loosely connected articles on a topic, while also offering all members of the research team the opportunity to co-author chapters. The research project and this edited volume have been made possible with the help of the following individuals (with their institutional affiliations at the time of the research).

Aga Khan Agency for Habitat (AKAH): Onno Ruhl, Nawab Khan, Javaid Ahmed, Ibrar Shah, Muhammad Ayoub, Nazeer Ahmed, Muhammad Iqbal, Jafar Ali, Rahat Baig, Mehar Aftab, Jibran Shafa, Syed Kazim Raza, Mukham Khan, Muhammad Saleem, Nafas Bib, Teresa Wilcke.

Aga Khan University: Ehsan Solaimani, Donald Dinwiddy, Amal Imad, Samantha Griffin, Irfan Haji, Anjum Halai, Mola Dad Shafa, Charlotte Whitting, Alex Khan, Lukhi Linnebank, Saima Ali, Kanika Lang, Gulru Azatshoeva, Alina Sabah.

Karakoram International University (KIU): Garee Khan, Irfan Ali.

Government of Gilgit-Baltistan (state departments): Anwar Jamal (GB House), Muhammad Iqbal (*former Minister of Works*), Usman Ahmed (Commissioner, Gilgit), Muhammad Naeem (former Director, LG&RDD), Sheraz Nadeem (Assistant Director, LG&RDD), Khadim Hussain (Assistant Director, Environment Protection Agency), Abdul Ali (Deputy Secretary, Public Works Department).

Enumerators: Sona Karim, Waseem Akram, Faryal Akhbar, Aziz Karim, Komal Nabi, Lubna Ahmed, Muhammad Khan, Muhammad Mujtaba Hussain, Saira Akber, Saira Bano, Samrina Nazar, Sayeed Murad Shah, Islam Nabi, Zahid Hussain, Masooma, Iftikhar Ali, Yasmeen.

The editors would also like to thank all water and sanitation committee (WSC) members and community members who were interviewed and participated in stakeholder consultations and focus group discussions.

Author biographies

Jeff Tan is Professor of Political Economy at Aga Khan University, Institute for the Study of Muslim Civilisations in London and was the principal investigator on the British Academy Urban Infrastructures for Well-Being grant that funded this research project.

Stephen Lyon is Professor of Anthropology and Dean of the Faculty of Arts & Sciences, Aga Khan University, in Pakistan, and was a co-investigator on the project.

Attaullah Shah is Professor of Civil Engineering and Vice Chancellor of Karakoram International University in Pakistan, and was a co-investigator on the project.

Anna Grieser is an anthropologist working on water in Gilgit and was a postdoctoral research fellow on the project.

Matt Birkinshaw is a geographer working on water and was a postdoctoral research fellow on the project.

Saleem Khan is Manager at the Aga Khan Agency for Habitat (AKAH) in Gilgit and was a research partner on the project.

Saleem Uddin was Community Development Officer/Social Organiser at AKAH in Gilgit, and a research coordinator on the project.

Fatima Islam is an Aga Khan University alumnus from Gilgit and was a research assistant on the project.

Yasmin Ansa is Health and Hygiene Coordinator at AKAH and was a research coordinator on the project.

Karamat Ali is Assistant Professor of Environmental Sciences at Karakoram International University in Gilgit and was a researcher on the project.

Sabrinisso Valdosh is an Aga Khan University alumnus and was a research assistant on the project.

Manzoor Ali is Associate Professor of Physics at Karakoram International University in Gilgit and was a researcher on the project.

Mushtaque Ahmed is Water Quality Management Coordinator at AKAH and was a research coordinator on the project.

Acronyms and abbreviations

ADP	Annual Development Program
AKAH	Aga Khan Agency for Habitat
AKDN	Aga Khan Development Network
AKHS	Aga Khan Health Services
AKPBS-P	Aga Khan Planning and Building Service Pakistan
AKRSP	Aga Khan Rural Support Programme
BoQ	bill of quantities
CapEx	capital expenditure
CapManEx	capital maintenance expenditure
CBWM	community-based water management
CHIP	Community Health Intervention Programme
CM	community management
DWSS	drinking water supply scheme
EPA	Environmental Protection Agency
FGD	focus group discussion
GB	Gilgit-Baltistan
GBLA	Gilgit-Baltistan Legislative Assembly
GLAAS	Global Annual Assessment of Sanitation and Drinking-Water
GLOF	glacial lake outburst flood
GoGB	Government of Gilgit-Baltistan
GoP	Government of Pakistan
H&H	health and hygiene
IAUP	Integrated Area Up-gradation Project
JSR	Joint Sector Review
KfW	German state-owned development bank
LB&RDD	Local Bodies and Rural Development Department
LG&RDD	Local Government & Rural Development Department
LGD	local government department
LHW	Lady Health Worker
MHVRA	multi-hazard vulnerability and risk assessment
MoCC	Ministry of Climate Change
NDWP	National Drinking Water Policy
O&M	operations and maintenance
OpEx	operating expenditure
PHED	Public Health Engineering Department
PPAF	Pakistan Poverty Alleviation Fund
PPP	public–private partnership
PWD	Public Works Department

SDGs	Sustainable Development Goals
SHIP	School Health Intervention Programme
SM	social mobilization
SOP	standard operating procedure
TDR	term deposit receipt
TDS	total dissolved solids
ToP	terms of partnership
VO	village organization
WASA	Water and Sanitation Agency
WASEP	Water and Sanitation Extension Programme
WASH	water, sanitation, and hygiene
WHO	World Health Organization
WO	women's organization
WSC	water and sanitation committee
WSHHSP	Water, Sanitation, Hygiene, and Health Study Project
WSI	water and sanitation implementer
WSO	water and sanitation operator
WSS	water supply and sanitation

CHAPTER 1

Introduction: Why community water management?

Jeff Tan

1.1 Introduction

Community-based water management (CBWM) is the leading model for implementing and sustaining (rural) drinking water supply services (DWSS) in low- and middle-income countries. Starting out in rural locations and often targeting remote, isolated, and sparsely populated villages that were excluded from piped water networks, CBWM has since been scaled up and extended into urban areas. This is despite the evidence of poorly or non-functioning systems as a result of inadequate maintenance because of weak community management and financial capacities. Lower income communities in particular are seen as lacking the necessary capacities to operate, maintain, and manage DWSS and are also often unable or unwilling to pay regular water tariffs necessary for operations and maintenance (O&M). This has led to a recent emphasis on the importance of creating an enabling environment for community management by strengthening community capacities through technical, financial, policy, and institutional support.

These conditions and requirements for CBWM have long been established (see e.g. McCommon et al. 1990; Schouten and Moriarty 2003) and their renewed and ongoing emphasis, most notably under the idea of 'community management plus' (CM+) (see e.g. Lockwood and Smits 2011; Hutchings et al. 2015), indicates that communities continue to struggle to manage, maintain, and, more critically, finance operations. This raises the question of whether the ongoing poor performance of CBWM is because the necessary external support has not been provided, as is the general consensus, or if there might be inherent limitations to the CBWM model itself, including assumptions that such support can be provided. In particular, this support places the state at the centre of sustainable CBWM in countries where state capacities are usually weak. At the same time, the idea of creating an enabling environment is based on the implicit recognition that the wider context matters for successful community management.

This edited volume looks at how these inherent limitations of the CBWM model impact sustainability and scalability, and how the wider social context may hinder or facilitate the application of CBWM principles of participation, ownership, control, and cost sharing. In particular, the ongoing lack of external support suggests some fundamental problems in the CBWM model related to: (a) a fragmented financing model that favours short-term project and programme funding over long-term sector-wide investment (see e.g. Harvey and Reed 2006); (b) the lack of state institutional capacity to support and finance CBWM; and (c) the irregular and non-payment of tariffs compounded by very low tariffs (to ensure affordability) that undermines cost recovery.

The case study of the Water and Sanitation Extension Programme (WASEP) implemented by the Aga Khan Agency for Habitat (AKAH) in northern Pakistan is especially pertinent and offers important lessons as a successful, large-scale implementation of CBWM. The success of WASEP can be defined and assessed in terms of: (a) the functionality of schemes; (b) the ability of communities to undertake O&M; (c) water infrastructure and water quality; and (d) sustained operations. Since its introduction in 1997, WASEP has delivered clean piped drinking water to 459 settlements covering 47,629 households with an estimated population of over 400,000 people. In terms of delivery and functionality, WASEP thus stands out as an 'island of success' in relation to the widespread failure of other CBWM schemes (see Davis and Iyer 2002, cited in Lockwood and Smits 2011: 76). It also demonstrates how the level of community participation, ownership, control, and cost sharing depends on the wider context and local history, including the different social compositions of (rural and urban) communities.

This edited volume asks whether the WASEP model of CBWM is sustainable and scalable to urban areas, and in doing so seeks to contribute to the wider debate on the sustainability and viability of CBWM. It draws from evidence from a large-scale household survey supplemented by interviews with community management committees, focus group discussions with households and management committees, an engineering audit of water infrastructure, and water quality tests. The significance of WASEP is not only that it is an example of successful implementation and sustained operations but that despite this it also faces many of the same challenges reported in much less successful and failed CBWM projects.

The chapters in this volume examine and locate these challenges within the inherent limitations of the CBWM model and wider social context. The volume asks how WASEP is affected by wider institutional and governance structures (including state support or the lack of) and local factors such as the socioeconomic context, sectarian composition, and cultural attitudes and practices. The chapters provide the wider context to examine WASEP's implementation of CBWM principles of participation (in general and by women) and cost sharing, but also its distinct focus on engineering design and infrastructure in contrast to the CBWM focus on governance. The following section provides an overview of the underlying principles of CBWM and conditions identified for sustainability. Section 1.3 provides an overview of the book's chapters and how these help illustrate the challenges in sustaining and scaling up CBWM.

1.2 How sustainable and scalable is community-based water management?

This section examines how the sustainability and principles of CBWM are invariably constrained by this model's inherent limitations. CBWM is based on the principles of participation, control, ownership, and cost sharing. Participation refers to an 'active process whereby beneficiaries influence the direction and execution of development projects rather than merely receive a share of project benefits' (Paul 1987: 2) and includes village-wide meetings and management or water user committees. Participation is the basis of a demand-driven or demand-responsive approach that is seen to better respond to community needs because communities express a demand from the start, identify 'the problem they wish to address and the level of services and technology they want and can afford' (McCommon et al. 1990: 33), and thus demonstrate a willingness to manage and pay for the water service.

Women's participation is increasingly emphasized as an important part of wider community participation and management in CBWM and is often a (donor) requirement. As (rural) women are responsible for collecting water, they stand to gain the most from water services and are therefore expected to support and participate in CBWM schemes. Women's participation in water management committees is also important as gender representation is said to enhance system functionality while promoting equitable access (Whaley and Cleaver 2017).

Ownership often loosely refers to a 'sense of ownership' (rather than legal ownership) that is usually seen as sufficient to facilitate community decision-making and control over the type of water services desired, with beneficiaries assisting and having a say in project planning, design, and implementation (McCommon et al. 1990: 7). There is disagreement about what constitutes actual ownership, with community control argued to require legal ownership essential to protect the source, prevent disputes, and resolve conflict (Schouten and Moriarty 2003). Ownership is important as it is seen to create a willingness to share project costs through (financial or in kind) contributions for capital costs, and regular tariff payments for recurring (operational) costs. A willingness to pay in turn demonstrates a willingness to accept responsibility for the water supply (Schouten 2006) and is also 'a measure of management potential' (McCommon et al. 1990: 14).

Community management can thus be defined as the community having 'responsibility, authority, and control over the development of such services' (McCommon et al. 1990: 2). It is often seen as involving 'routine operational duties such as record keeping, accounting, and payment collecting under a system predefined by an external agency' (McCommon et al. 1990: 10). The essence of community management though is said to lie in control rather than O&M because 'control covers the decision-making powers that put a community truly in charge' including 'the ability to make strategic decisions about how a system is designed, implemented, and managed' by selecting service levels, setting tariffs, and undertaking (or contracting out) O&M (Schouten and Moriarty 2003: 165).

This is reflected in variants of the community-based management model (World Bank 2017). The most common CBWM variant are informal community organizations that manage O&M but are not legally recognized as service providers because they are not registered or because of the absence of government policy or legislation to facilitate legalization. A second variant are formal community service providers that perform the same tasks but are legally recognized as service providers thereby qualifying for government support, at least in theory. This type of community provider may contract out certain O&M tasks to individuals (e.g. plumber or scheme attendant) or the entire O&M to a private operator who receives remuneration through the sale of water, with the community organization exercising oversight (World Bank 2017: 100).

Cost sharing between government, donors, and users is a central principle of community management that enables communities to 'express their demand for services at least partially through contributions to the investment costs' thereby promoting a sense of ownership and financial sustainability (Lockwood and Smits 2011: 113; Hope et al. 2020). Community participation in, and 'ownership' of, drinking water supply systems, it is argued, thus ensures that water services are sustainable through the regular collection of tariffs for ongoing O&M. Women's participation can be expected to improve sustainability by increasing broader support, a sense of ownership and control, and payment of tariffs for O&M. The central thesis of CBWM is that 'strong community management leads to sustainable water supply and sanitation systems' (McCommon et al. 1990: 3) and effective CBWM is a precondition for sustainable water systems (Lockwood and Smits 2011; Mugumya et al. 2015; Naiga 2018), with water user committees being a key component. The underlying premise is that (rural) communities are willing to pay for water services given missing, inadequate, or unsatisfactory public service delivery, and that these communities have the resources and capacity, and know what is best for them.

However, high levels of breakdown rates and widespread non-functionality, especially in sub-Saharan Africa and India (Churchill 1987; Baumann 2006; Reddy et al. 2010; Lockwood and Smits 2011), with 'one in four rural waterpoints in rural Africa broken at any one time' (Hope et al. 2020: 174), illustrate the lack of sustainability (typically defined in terms of 'functionality' and measured by the percentage of working water points) and continued operation of water services (Lockwood and Smits 2011; Carter 2021). This has recently been expanded to include indefinite operations, defined as 'secure water supply not for the lifetime of a system but for the lifetime of a community' (Schouten and Moriarty 2003: ix). Sustainability is thus tied with the availability of financing to cover all recurring costs related to ongoing operations as well as all future costs associated with asset renewal (refurbishment) and service expansion.

The lack of sustainability in CBWM projects has been blamed on internal and external factors. Internal factors relate to the social context and community limitations due to unfavourable conditions such as low initiative, weak leadership, and insufficient capacity. External factors refer to an insufficient institutional framework including policy and legislation, and financial, material, technical, and management support from the state, donors, and NGOs (Lockwood 2004). A common thread running through the CBWM literature is the importance of external or post-construction support from government, NGOs, and the private sector that includes financial or technical support for hardware or management rehabilitation (see e.g. McCommon 1990; Schouten and Moriarty 2003; Lockwood 2004; Lockwood and Smits 2011; Hutchings et al. 2015, 2016; Kelly et al. 2017; Miller et al. 2019; Carter 2021). The emphasis here is on the 'software' or governance component of water systems (as opposed to 'hardware' or physical infrastructure) and the need to create an 'enabling environment' to overcome these internal (community) and external (institutional) constraints. This 'governance-analysis' that is also part of CM+ seeks to strengthen community capacity through technical, financial, and institutional support.

However, the requirements of CM+ are nothing new and were always part of the original CBWM model, where preconditions included: the appropriate level of technology, community willingness and capacities, policy support, and external support (technical, training, financial) (McCommon et al. 1990: 11-12; Schouten and Moriarty 2003). Subsequent preconditions included social conditions such as community cohesion and the absence of conflict (Schouten and Moriarty 2003). Policy support includes legislation to clearly define community ownership and authority, and support for the enforcement of tariff collection as well as sanctions for non-payment. Policy will be needed for harmonization and coordination among development partners and government, and alignment with national policies and systems to promote sector-wide financing of water services, and a system-wide approach (World Bank 2017). Life-cycle costing is required to factor in construction, maintenance, replacement, and indirect support costs (including state capacity-building, planning, and monitoring at district and national levels) to ensure the delivery of water services 'not just for a few years, but indefinitely' (Fonseca et al. 2011: 6).

Although the importance of ongoing external support was identified as early as 1990 in a World Bank report (McCommon et al. 1990), post-construction support and water and sanitation services have continued to be consistently under-funded (Lockwood and Smits 2011; UN Water n.d.). The early promotion in the 1990s of CBWM in Pakistan included community contributions to capital cost and taking over responsibility for O&M to free up resources to invest in bridging the piped water access gap, but 'a lack of support to communities for financial and technical management of schemes meant that a number of schemes became dysfunctional, requiring reinvestment' (World Bank 2016: 11). A more recent World Bank study of 16 countries in Asia, Africa, and Latin America found that only 17% of CBWM projects received external support (World Bank 2017).

This ongoing inability to properly support and finance CBWM raises two immediate issues. The first is why the necessary external support has not been provided to communities given widespread knowledge of the preconditions for successful CBWM. The reasons for this can be found in two inherent limitations of the CBWM model: a short-term, fragmented financing model and the lack of state institutional capacity. The current CBWM model of financing is fragmented, favours short-term project and programme funding over long-term, sector-wide investment (see e.g. Harvey and Reed 2006), and trades sustainability for coverage, with funding not covering all costs (Lockwood and Smits 2011) let alone the cost of external support and institutional building.

Furthermore, governments and donors do not know the real costs of the entire life cycle at local and national levels and cannot specify financing requirements or determine who should finance these (Lockwood and Smits 2011). The requirement for external support also places additional institutional and financial demands on a fragmented, weakened, and fiscally constrained state, where there is no clarity on national guidelines, and ambiguous, conflicting, and contradictory asset ownership, and with many governments in developing countries unable to harmonize and align policies and legislation (World Bank 2017).

The second issue that arises is whether this support would promote sustainability given a third limitation of the CBWM model: the tension between affordability and cost recovery. CBWM in low-income and especially rural communities is often characterized by irregular and non-payment of tariffs (Schouten and Moriarty 2003; Mugumya 2013; World Bank 2017; Hope et al. 2020) with tariff levels having to be set too low to cover operating costs. Very low- to low-income households are not only less able or willing to pay but also less likely to participate (see e.g. McCommon et al. 1990; Hutchings et al. 2016), with community participation in CBWM activities 'more worrying if water users live in poor and marginalized rural areas' (Mugumya 2013: 74). Additionally, women's participation is also low, especially on water user committees (see e.g. Hannah et al. 2021) because the requirements for this are usually part of wider (and often external) gender inclusivity and women's empowerment agendas that may conflict with existing (local) norms especially in more socially conservative communities.

Sustainability thus cannot be restricted to just the community management component of CBWM because the capacity of communities to manage and contribute to (operational) cost recovery also depends on the overall financing of water services. The largest expense in the delivery of water services is the cost of water infrastructure, which accounts for around 50% of expenditure, followed by capital maintenance expenditure or asset renewal and replacement at 25% (Carter 2021). The long-term financing of water infrastructure is thus central not just for O&M (e.g. by developing community management capacities) and hence ongoing functionality, but also in terms of the initial engineering and infrastructure, its renewal or rehabilitation, and its extension as part of service expansion to meet the demands of growing populations. Sustainability thus entails maintaining existing water systems and 'scaling up in time' (for improved, sustainable services), which is a condition for increasing coverage to ensure no one is excluded by 'scaling up in space', to deliver water supply services indefinitely (Schouten and Moriarty 2003: 289).

1.3 Why WASEP?

The following chapters examine how the sustainability and scalability of WASEP are shaped by the wider social, cultural, and institutional context, and constrained by the inherent limitations of the CBWM model. WASEP is implemented by AKAH, an agency of the wider Aga Khan Development Network (AKDN), a network of NGOs located mainly in the Global South.

Chapter 2 provides the background and context needed to frame the WASEP case study, beginning with the mountain province of Gilgit-Baltistan where climate and geological hazards present specific engineering challenges that help explain WASEP's emphasis on engineering design, and where sectarian and (old and new) settler composition affect the levels of unity (and conflict), cooperation, and hence participation. It provides an overview of WASEP's history, the types of water-related projects undertaken, and funding sources, and details of the research project behind this edited volume, including the methodology and site selection. The data on the total 459 WASEP projects demonstrates the scale of the programme and helps locate the 50 sample sites across the 10 districts of the province. It illustrates how funding is based on the CBWM financing model that is largely reliant on external funding sources and is focused on specific phases of delivery, each with a target number of projects. Finally, it highlights the key differences between rural and urban projects in terms of community composition and community share of total project costs. These help situate the subsequent chapters on participation, conflict, cost sharing, and community responses to damaged water infrastructure.

Chapter 3 examines the problems of water governance in Pakistan, which is characterized by weak state and institutional capacities, and institutional fragmentation in common with other lower middle-income countries, with multiple authorities and overlapping responsibilities and jurisdictions on federal and local levels. Effective water governance and management are undermined by the absence of exclusive legislation at the federal and provincial levels, and the involvement of national institutions is limited because the responsibility for service delivery has been transferred to provincial governments and local institutions as part of decentralization policies. This is mirrored in Gilgit-Baltistan where water services are provided by a (growing) number of (overlapping) state departments with a lack of coordination and clarity about responsibility and support for communities, raising questions about whether the state has the technical and political capacity, for example, to legislate on traditional water rights and asset ownership.

Chapter 4 introduces the WASEP model and identifies the factors behind its successful implementation and the ongoing functionality of WASEP projects. WASEP was created as a response to earlier unsuccessful attempts that relied solely on the delivery of DWSS to reduce waterborne diseases. WASEP's integrated approach is built on the CBWM principles of participation, ownership, control, and cost sharing, as well as health and hygiene education to promote behavioural change, while retaining AKAH's focus on engineering solutions and water quality management. This emphasis on 'hardware' or water infrastructure is distinct from the focus of much of the CBWM literature on 'software' or governance and management. Social mobilization is central to WASEP's promotion of participation, ownership, control, cost sharing, and health and hygiene education, and unlike recommendations in the literature on community control, AKAH plays a leading role as the implementing NGO in the design and delivery of water infrastructure.

Chapter 5 looks at how the wider social and historical context is important in explaining successful rural participation in WASEP. Evidence from the household survey shows how participation varies depending on the social context. Participation is higher in socially more homogeneous rural communities that have also benefited from grassroots work by various AKDN agencies over the previous decades. However, actual participation levels are lower than the demand (or desire) indicated by households for water services, and low at the management level and in urban projects, consistent with the wider evidence. More critically, high rural demand and general participation do not translate into the ability or willingness of households to pay that is key to cost sharing and sustainability. In contrast, urban households have much lower levels of participation but a greater willingness or ability to pay regular tariffs, suggesting that differences in rural and urban household income levels and water sources matter. The evidence also raises questions about whether participation or the willingness and ability to pay are more important for sustainability.

Chapter 6 examines the role of women's participation in WASEP. Women are seen as central in CBWM as the main intended beneficiaries of (rural) water services and given their roles as carers who usually bear the responsibility of health, sanitation, and hygiene for the household. Women in WASEP are expected to participate and support CBWM, and to play a leading role in health and hygiene awareness raising through the (paid) position of a water and sanitation implementer (WSI) that is financed through monthly tariffs. The evidence shows that women's participation is associated with more regular tariff payments, but that overall participation levels are low, and very low in urban households and at the management level, with the mandatory position of (woman) WSI unfilled in many projects. The level of women's participation also varies depending on the social context, specifically local customs and norms.

Chapter 7 looks at water-related conflict and conflict management in WASEP projects as a basic requirement for successful community management. In the context of Gilgit-Baltistan this conflict can manifest itself along sectarian lines that define communities, or between old and new settlers in urban areas over traditional water rights and access to water sources. WASEP terms of partnership seek to minimize conflict by making community participation and hence unity a prerequisite. AKAH's ability to manage conflict also builds on a long history of AKDN agencies working with communities in the region. Nonetheless conflict invariably occurs at different phases of a project and was not surprisingly higher and lasted longer in urban projects with more diverse communities, with the main sources of conflict for both rural and urban sites being over resource distribution and tariff payments.

Chapter 8 looks at the tension between affordability and cost recovery and how this has impacted the financial performance of WASEP projects. The evidence shows that, despite higher participation, rural households are the least likely to pay regular tariffs and most opposed to the payment for water. Irregular and non-payment of tariffs in many rural projects, along with very low tariff levels, has meant that preventive O&M is replaced with ad hoc household payments as and when (minor) repairs are needed. This arrangement has kept most rural projects functioning but has also needed external financial or material support for major repairs that was not always available. As a result, only a quarter of the WASEP rural sample are estimated to have an operating surplus, and only a third were able to increase the value of their endowment funds for major repairs, meaning that community funds may not be available for major repairs in future. Urban WASEP projects are financially more sustainable given the higher tariff levels and a greater willingness and ability to pay despite higher operating costs although regular tariff payments also vary between different urban locations.

Chapter 9 examines the importance of WASEP's ongoing focus on engineering design and its impact on water infrastructure and water quality. It looks at AKAH's engineering design as the foundation for the delivery of clean drinking water and how AKAH retains control over all engineering and technical decisions contrary to the CBWM principle of community control, including decisions on system design and implementation. AKAH's engineering and technical expertise is reflected in higher engineering audit scores for water infrastructure and water quality compared to control sites, with water quality meeting WHO standards for drinking water for almost all WASEP projects. This is itself testament to the importance of 'hardware' or physical infrastructure as the basis for the delivery of clean drinking water and for long-term functionality.

Chapter 10 looks at how WASEP meets the engineering challenges of delivering water services in an inhospitable mountainous region that characterizes Gilgit-Baltistan, where high altitude, difficult terrain, and low temperatures, along with natural hazards, pose significant technical and practical challenges. In order to secure clean water sources, WASEP projects have on average longer pipe networks that are also buried below the frost line to prevent freezing and, aside from two neighbouring urban projects, have had fewer days lost to damage and disruption from natural hazards, in particular floods, compared to control sites. As engineering design can mitigate but not prevent the impact of natural hazards, community responses are important for repairing damaged infrastructure and restoring water services. The second part of this chapter examines the importance of these responses, mainly through self-help initiatives including additional financial contributions, with community management capacity important in social mobilization. The evidence also suggests that external support is important to restore water services after damage from natural hazards, with over half of rural and all urban management committees seeking and receiving external support from AKAH or government agencies.

Chapter 11 brings together the findings in this volume to reappraise the sustainability and scalability of WASEP. It shows how the successful implementation of WASEP marks it out from the wider failure of CBWM and highlights the key features of WASEP related to the wider socioeconomic context and a continued emphasis on engineering. High levels of community participation and (financial) contributions in WASEP in particular are unique to the region and its history, including the earlier work of previous AKDN agencies, and suggests this may not be easily replicable in other contexts. Nonetheless, WASEP's experience also reinforces the inherent limitations of the CBWM model due to the inability or unwillingness of (poor) households to pay regular tariffs, compounded by tariff levels that are often too low to cover costs, and a fragmented financing model that favours short-term projects with no allocation of funds for major repairs, system rehabilitation, and service expansion necessary for sustainability. As a result, while the WASEP model offers potential lessons to improve CBWM through examples of best practices, sustainability and scalability will remain elusive without a fundamental re-evaluation of the viability of the CBWM model itself.

CHAPTER 2

Background: Gilgit-Baltistan, WASEP, and the research

Jeff Tan, Anna Grieser, and Matt Birkinshaw

2.1 Introduction

This chapter introduces the region of this research (Gilgit-Baltistan in north Pakistan), the community-based water management (CBWM) case study (the Water and Sanitation Extension Programme or WASEP), and the selection of the research sites. It provides the necessary background and context, identifies some of the distinct challenges, and highlights the key features of CBWM discussed in Chapter 1 as the basis for more detailed analyses in the subsequent chapters. The case study of WASEP is based on a research project on 'scaling up and transferring community-managed rural water systems to urban settings' funded by the British Academy under its 2019 Urban Infrastructures for Well-Being programme. The programme aimed to 'address the challenges of generating and maintaining well-being in the context of rapid urbanisation and infrastructure development in cities of the Global South' (British Academy 2019: 2).

The research was first suggested by the Aga Khan Agency for Habitat (AKAH), the implementing NGO, to determine if WASEP is sustainable and scalable to urban centres. It involved an interdisciplinary collaboration between scholars from the social sciences (political economy, anthropology, geography) at the Aga Khan University-Institute for the Study of Muslim Civilisations (AKU-ISMC) in London, and engineering sciences (engineering, environmental sciences) from Karakoram International University (KIU) in Gilgit-Baltistan, in partnership with engineers and development practitioners from AKAH.

While both AKU-ISMC (the lead institution) and AKAH are part of the broader Aga Khan Development Network (AKDN), the aim of the research was not to support, promote, or justify WASEP but rather to critically assess its sustainability in order to inform future plans, and in doing so, contribute to the literature on CBWM. It is also hoped that some of the extensive but scattered data on WASEP can be brought together in one place for researchers, development practitioners, country experts, and the general reader.

Section 2.2 provides an overview of Gilgit-Baltistan in terms of its location, geographical features, administrative structure, and social composition necessary to inform the sample selection. Section 2.3 introduces funding sources as the basis of the sample selection. Section 2.4 discusses the research methodology, including sample selection, and summarizes some of the main research findings for analysis and discussion in the subsequent chapters.

2.2 Gilgit-Baltistan: geography, administration, population

Gilgit-Baltistan is located in the mountainous region of north Pakistan bordering Afghanistan, China, and Indian-administered Kashmir, and along three mountain ranges – the Hindu Kush, Karakoram, and Himalayas (Figure 2.1). With the highest peaks at over 8,000 m (including K2, the world's second highest peak) and around 700 peaks above 6,000 m (Sharma et al. 2019; Ahmad et al. 2020), Gilgit-Baltistan has been described as a vertical or cold desert, with the valleys only receiving around 100–200 mm of precipitation annually (Archer and Fowler 2004). Precipitation at altitudes above 3,000 m can be as high as 2,000 mm per square metre per year, forming year-round snow and glacial fields, especially at altitudes above 6,000 m. Gilgit City, the provincial capital located in Gilgit district at an altitude of 1,490 m, receives around 130 mm of precipitation annually, with temperatures ranging from –10° to 40° Celsius (JICA/GoGB 2010). As glacial melt running into streams and rivers is the main source of water in Gilgit-Baltistan, the availability of water varies greatly between the winter and summer months. The region is also vulnerable to geological hazards such as avalanches, glacial lake outburst floods, and landslides, which create additional challenges for water infrastructure.

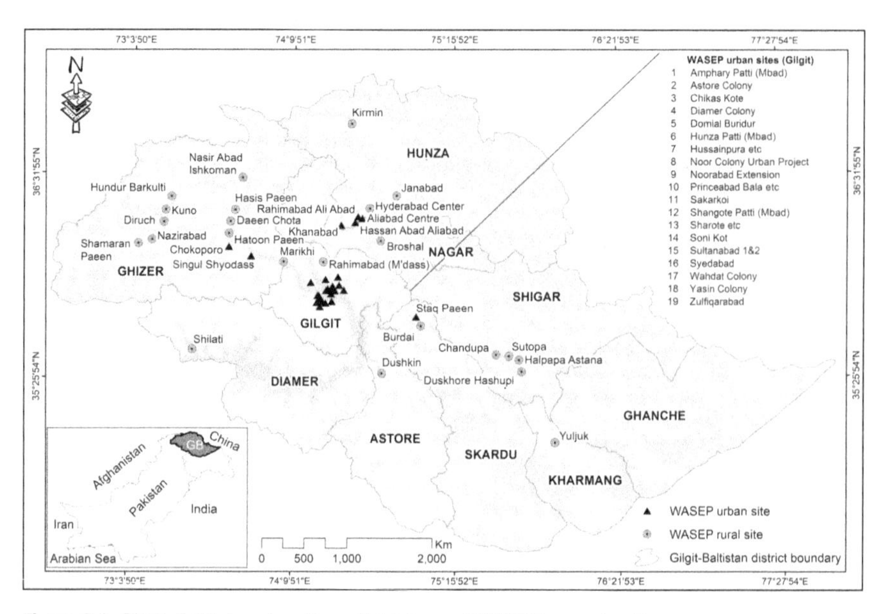

Figure 2.1 Gilgit-Baltistan: location, districts, and WASEP sample sites
Source: map by Karamat Ali and Garee Khan

Table 2.1 Gilgit-Baltistan: administrative structure, 2019

Semi-autonomous province	*Division*	*Divisional capital*	*District*	*District capital*
Gilgit-Baltistan	Gilgit	Gilgit	Gilgit	Gilgit (City)
			Ghizer	Gahkuch
			Hunza	Aliabad
			Nagar	Nagar
	Baltistan	Skardu	Ghanche	Khaplu
			Kharmang	Kharmang
			Shigar	Shigar
			Skardu	Skardu
	Diamer	Chilas	Diamer	Chilas
			Astore	Eidghah

The province is semi-autonomous and was previously part of Pakistan's northern regions (along with Azad Jammu and Kashmir). It remains constitutionally undefined as a disputed territory awaiting a United Nations plebiscite to decide the fate of Kashmir (Holden 2019: 3). Administratively, Gilgit-Baltistan is divided into three divisions and a number of districts, with each district consisting of sub-districts (*tehsils*), union councils, and local councils (Table 2.1). The number of districts has steadily increased over time from 7 (Astore, Diamer, Ghanche, Ghizer, Gilgit, Hunza-Nagar, Skardu) to 10 during the research period (2019–2021) (Figure 2.1 and Table 2.1) and currently stands at 14. For the purposes of this research, the 10 official districts from 2019 were used – Astore, Diamer, Ghanche, Ghizer, Gilgit, Hunza, Kharmang, Nagar, Shigar, and Skardu (Table 2.2). To avoid confusion, Gilgit (the district capital of Gilgit district and main research site) is referred to here as Gilgit City.

The population of Gilgit-Baltistan was estimated at 1.49 million in 2017 (unpublished 2017 census) (see Table 2.2) across an area of around 72,000 km^2. Given the mountainous topography, most settlements are located in valleys at altitudes between 1,000 m and 3,000 m, with seven major valleys – Astore, Diamer, Ghanche, Ghizer, Gilgit, Hunza-Nagar, and Skardu – also constituting administrative districts. Unlike other regions in Pakistan, the majority of the population in Gilgit-Baltistan is Shia, with large minorities of Sunni and Ismaili Muslims (Grieser and Sökefeld 2015; Hunzai 2013). This diversity is reflected in the seven local languages spoken in the region and settlements based on language, sectarian identity, kinship, and quasi-kinship (see e.g. Grieser and Sökefeld 2015), particularly in the countryside but also in urban centres where (newer) settlements are sometimes named after the original districts or places of origin (e.g. Astore Colony and Diamer Colony in Gilgit City). The sectarian and settler composition of communities are an important part of CBWM as these potentially affect unity, participation, cooperation, and conflict (discussed in the following section and summarized in Tables 2.4 and 2.5).

Table 2.2 WASEP projects by district, Gilgit-Baltistan as at January 2021

District	*Households (district)*	*Population (district)*	*WASEP projects*	*Households (WASEP)*	*Population (WASEP)*	*Sample sites*	*Households (sample)*	*Population (sample)*	*Household responses*
Astore	12,135	95,416	17	1,313	12,798	1	203	1,861	60
Diamer	31,063	269,722	6	402	5,402	1	46	582	32
Ghanche	20,995	156,697	25	1,883	17,154	0	0	0	0
Ghizer	23,345	172,696	188	13,620	119,809	14	899	7,500	614
Gilgit	40,156	285,236	50	16,129	131,758	20	10,110	82,650	1,703
Hunza	8,241	51,372	65	6,218	51,448	6	1,769	13,747	369
Kharmang	7,788	54,613	1	50	442	0	0	0	0
Nagar	10,839	71,746	15	1,413	12,554	1	81	652	58
Shigar	9,725	74,540	26	2,398	21,521	5	431	4,505	162
Skardu	32,139	260,836	66	4,203	39,537	2	155	1,156	134
Total	196,426	1,492,874	459	47,629	412,423	50	13,694	112,653	3,132

Note: Households and population data from 'Provincial Results Census – 2017 of Gilgit-Baltistan'.

Gilgit City is the capital and the largest urban centre in Gilgit-Baltistan with around 120,000 inhabitants. It is characterized by a piecemeal, fragmented water supply, especially in newer settlements where many residents are migrants from the surrounding valleys without access (water rights) to the two main freshwater streams supplying Gilgit City (Grieser 2018). As a result, the neighbourhood of Jutial in Gilgit City has been affected by water shortages, with new residents organizing additional mechanized water supply systems drawing from river water with the help of the Aga Khan Rural Support Programme (AKRSP) and later under WASEP. Mechanized water systems introduce an additional set of problems because electricity supply is unreliable and not always available throughout the day. This means that even if clean water is available, it is not possible to get this to households during daily, and often prolonged, power cuts.

2.3 WASEP overview: projects and funders

WASEP is currently implemented by AKAH, a more recent agency of the AKDN created in 2016 as part of organizational restructuring that brought together several agencies and programmes working on habitat and disaster preparedness and relief since the 1990s. This included Aga Khan Planning and Building Services (AKPBS) that also previously administered WASEP. Water projects have also been delivered by AKRSP. WASEP thus has a much longer history than AKAH and is often referred to as both a programme and an entity. In this volume, WASEP refers to the programme that delivers drinking water supply schemes (DWSS) while AKAH refers to the implementing agency, even if most WASEP schemes were previously implemented by earlier AKDN agencies before the creation of AKAH.

There are four types of WASEP projects: DWSS, sanitation only schemes, rehabilitation DWSS, and schemes under the Integrated Area Up-gradation Project (IAUP). WASEP projects are funded in phases depending on when funds are secured. The distribution of WASEP projects is not related to household numbers or population size of districts (Table 2.2) but a number of other criteria discussed in Chapter 4, most notably the desire of communities for water services and their willingness to participate and contribute. The significance of WASEP as a model of CBWM is reflected in its reach, accounting for almost a quarter of households in Gilgit-Baltistan (Table 2.2).

The first WASEP projects were implemented in 1997 with four rural DWSS funded by KfW (the German state-owned development bank) in Gilgit, Hunza, and Skardu. Table 2.3 provides an overview of the funding phases, number of WASEP projects, and beneficiary population between 1997 and 2021, along with the number of WASEP projects in the research sample. Funding phases are named after the main funding bodies or programmes: KfW (KfW Development Bank), PPAF (Pakistan Poverty Alleviation Fund), ECD (Early Childhood Development programme), EC (European Commission), JCVF (Japan Counter Value Fund), USAID (United States Agency for International Development), 7PV (Seven Priority Valleys Programme),

Table 2.3 WASEP: funding phases, drinking water supply schemes (1997–2021)

Funding phase	*Period*	*Projects*	*Households*	*Population*	*Rural sample*	*Urban sample*
KfW-I	1997–2001	67	7,277	62,876	6	0
Others	2001–2003	6	589	4,851	0	0
PPAF-I	2002–2004	11	593	5,398	2	0
PPAF-II	2004–2006	52	2,239	20,804	2	0
PPAF-V	2007–2008	50	1,742	19,648	3	0
ECD	2008–2010	3	404	3,636	0	0
PPAF-VI	2009–2009	14	448	4,648	0	0
EC	2010–2012	26	2,249	17,422	2	0
PPAF-VII	2010–2012	24	1,830	16,707	1	0
KfW-II	2010–2014	78	9,358	82,255	5	8
JCVF	2013–2016	72	6,220	55,298	2	3
USAID	2015–2017	1	485	3,880	0	0
7PV	2015–2018	23	1,565	13,316	1	0
GoGB*	2016–2021	25	12,390	99,120	0	14
PATRIP	2018–2018	7	240	2,132	1	0
Total	**1997–2021**	**459**	**47,629**	**411,991**	**25**	**25**

Note: KfW (KfW Development Bank), PPAF (Pakistan Poverty Alleviation Fund), ECD (Early Childhood Development programme), EC (European Commission), JCVF (Japan Counter Value Fund), USAID (United States Agency for International Development), 7PV (Seven Priority Valleys Programme), GoGB (Government of Gilgit-Baltistan), and PATRIP (PATRIP Foundation)

*Note**: GoGB funding is officially recorded by AKAH as two urban projects (Jutial and Danyore) but each location has multiple schemes, each represented by its own water and sanitation committee (WSC).

GoGB (Government of Gilgit-Baltistan), and PATRIP (PATRIP Foundation) (see Chapter 8 for more detail).

Funding for WASEP has historically focused on delivering DWSS to rural communities previously without access to safe drinking water. This is reflected in rural projects accounting for around 96% of DWSS in Gilgit-Baltistan in 1997–2021. Urban projects were only introduced much later, with six projects in Gilgit and Hunza funded by KfW (2011–2012, 2012–2013, and 2014) that included three urban IAUPs; two projects in Gilgit under JCVF (2012–2013); and most recently, 25 projects in Gilgit City under GoGB (2016–2021) (Table 2.3). This research only examines WASEP DWSS, along with two IAUP projects (because of their significance), across 8 out of 10 districts in Gilgit-Baltistan (listed in Table 2.2) – Astore, Diamer, Ghizer, Gilgit, Hunza, Nagar, Shigar, and Skardu – with projects in Ghanche and Kharmang excluded due to remoteness and inaccessibility. Table 2.2 includes the distribution of sample projects by district, total household and population numbers in the sample projects, and total responses from the household survey. Table 2.3 also lists the distribution of sample projects by funding phase and according to rural and urban projects.

2.4 Research and research sites

The aim of the research is to examine the success and sustainability of the WASEP model of CBWM and its scalability and transfer to urban centres by looking at three related topics: (1) project performance; (2) household characteristics, participation, and impact; and (3) engineering variables and performance. Success can be defined and assessed in terms of: (a) the functionality of schemes; (b) the ability of water and sanitation committees (WSCs) to undertake O&M; (c) water infrastructure and water quality; (d) sustained operations; and (e) indefinite operations. Sustainability ultimately refers to sustained operations indefinitely or for the lifetime of a community. Project performance covers functionality, O&M, and water quality. Household and engineering data are necessary to determine how these impact project performance and the sustainability and scalability of the WASEP model. AKAH supplied data on rural and urban DWSS in Gilgit-Baltistan that was collated, merged, and cleaned, providing a 'population' size of 459 DWSS (that included two urban IAUPs). WASEP projects and WASEP project participant households were chosen as primary units for sampling and analysis.

It should be noted that the number of projects in the AKAH datasets is defined in terms of funding. Hence the same site with funding over two different phases is counted as two projects even though it only involves one site. The same village could also be divided into different parts under the same funding source and phase but counted as different projects. For this research, a project is understood as having one WSC and one funding allocation, meaning that if the same village or same WSC is funded twice or by two different funds, this is understood as two different projects. Where WSCs in adjacent neighbourhoods (*mohallah*) have merged, these will still be counted as separate projects as was the case of Chokoporo, Domial Buridur, and Khanabad (in Gahkuch Bala), and Aliabad Centre and Rahimabad Aliabad that shared management under the same WSC. Projects, not villages/places, were chosen as a unit of analysis because villages contain multiple projects, and the study aims to understand socioeconomic factors behind the differences in project and WSC performance. Households were chosen as a secondary unit for sampling and analysis to understand the demographic differences across WASEP communities and effects on participant households. Schemes will be distinguished from projects, with the former referring to a group of projects.

The sample was first stratified into two groups, rural and urban, to understand the differences between rural and urban WASEP projects. In order to assess scalability, it was necessary to include as many urban projects as possible in the sample, even though there are only 42 projects categorized under 'urban DWSS', accounting for just 3.8% of total WASEP DWSS in AKAH's datasets. Funding under GoGB is officially recorded as two (umbrella) schemes by AKAH, covering Jutial and Danyore in and around Gilgit City. These comprise 11 individual projects with 11 separate WSCs in Jutial, and 14 projects and 14 WSCs in Danyore (discussion below and in Tables 2.6 and 2.7). As such, funding by GoGB is counted as 25 projects in this research. To stay within budget and to also obtain sufficient respondent numbers for random sampling to be a valid proportion of total households at the project level, the rural and urban samples were set at 25 projects each.

As financial performance is a core requirement for overall sustainability, the research assesses the financing of WASEP in terms of costs and revenue, based on calculations and estimates from financial data provided by AKAH and WSCs. Costs provided are in Pakistan rupees (PKR) and the conversion to US dollars (US$) is based on average annual exchange rates for each funding period or as otherwise specified. The choice of exchange rates is important in assessing the financial performance of WASEP because of the massive depreciation of the Pakistani rupee between 1997 (PKR41.1 = US$1) when the first WASEP project was implemented and 2021 (PKR163 = US$1) at the conclusion of this research project (see Figure 2.2).

To account for demographic difference, the samples were stratified by district. As covering all 10 districts of Gilgit-Baltistan would generate too many groups to further stratify within budget, the 10 districts were amalgamated into six 'district groups' on the basis of historical district areas as well as ethnic, linguistic, religious, and geographic similarities and differences: (1) Astore (as Shia); (2) Diamer (including Chilas, Darel, Tangir as Sunni); (3) Baltistan (including Ghanche, Kharmang, Shigar, Skardu, Rondu as Shia and geographically similar); (4) Hunza-Nagar (including

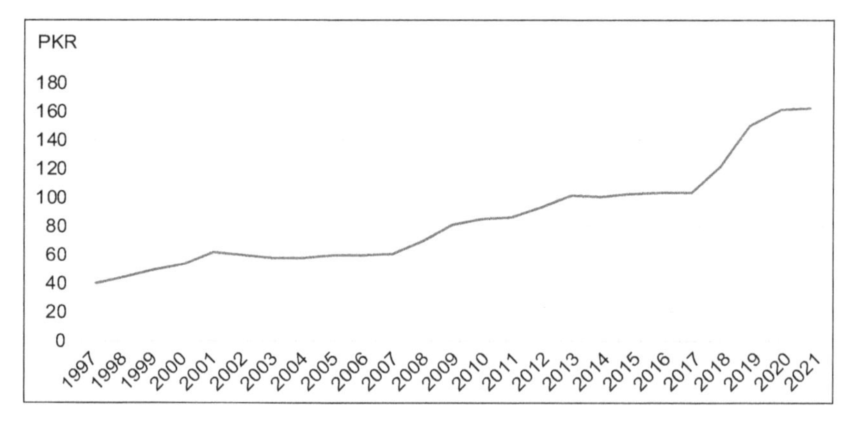

Figure 2.2 Average annual exchange rate: US$1 to PKR (Pakistani rupee), 1997–2021

Hunza and Nagar as geographically similar to each other and different from Baltistan despite religious difference); (5) Gilgit; and (6) Ghizer (including Gupis-Yasin). Data was then disaggregated into eight districts as described in Section 2.3 (see also Table 2.2).

The rural sample was divided proportionally to the distribution of rural projects across the six district groups, and rural projects were randomly sampled in these proportions. Nine of these rural projects were replaced due to the lack of cell (mobile) phone signal or village submergence in recent floods. The urban sample was stratified by district group proportional to project distribution, with urban projects only existing in Gilgit City, Aliabad in Hunza-Nagar, and Gahkuch in Ghizer. The Gilgit urban sample was then stratified to include a quota of single-district origin neighbourhoods as well as the range of majority/minority relations among different religious denominations. Tables 2.4 and 2.5 provide an overview of the sample sites in this research with key information that is discussed in detail in the following chapters. Figure 2.1 also provides a visual distribution of the rural and urban samples across the 10 districts of Gilgit-Baltistan.

The low number of urban projects is a reflection of the early focus of funding agencies (and the CBWM literature) on rural DWSS, with AKAH even having to classify some peri-urban projects as 'rural' to secure funding. This started to change in 2010 with the inclusion of two urban settlements of Gilgit City, one of Aliabad in Hunza district, and three on the fringes of Gahkuch in Ghizer district. Jutial is the largest new settlement in Gilgit City, which started out as a village on the outskirts of the city, grew rapidly, and was subsequently absorbed into the city boundaries. As mentioned previously, the Jutial scheme covers 11 individual projects as represented by separate WSCs for the different *mohallah*, in addition to two earlier WASEP schemes under KfW Phase II (Table 2.6). The urban sample comprises five GoGB-funded Jutial projects (Astore Colony, Diamer Colony, Wahdat Colony, Yasin Colony, Zulfiqarabad) and both KfW-funded Jutial projects (Noor Colony Urban Project, Noorabad Extension), covering 2,010 households. Table 2.6 also provides information on community composition and settler status where 'homogeneous' refers to projects where 90% or more of households are from one sect, and 'old settlers' refers to projects where 90% or more of households have lived in a settlement for over 50 years.

Danyore, along with neighbouring Sultanabad and Muhammadabad, was another former village on the outskirts of Gilgit City. The Danyore scheme comprises 14 individual projects and 14 WSCs (Table 2.7). However, half of Danyore WSCs cover multiple *mohallah* and as a consequence 13 of these WSCs (aside from Syedabad) are significantly larger in terms of total households (compared to an average of 78 households in village projects, and 250–350 households in the five 'pilot' urban projects), with heterogeneous populations in all WSCs.

Table 2.4 WASEP rural sample overview

Settlement	*District*	*Funding*	*Period*	*Operation (years)*	*Total households*	*Total population*	*Community composition*	*Community share %*	*System type*	*Source*	*Pipe length m*
Broshal*	Nagar	KfW-I	1998–98	23	81	652	Homogeneous	34.4	Gravity	Spring	7,250
Burdai	Skardu	PPAF-I	2002–03	18	34	272	Homogeneous	29.4	Gravity	Spring	3,068
Chandupa*	Skardu	KfW-I	2000–00	11	93	692	Homogeneous	39.1	Gravity	Spring	9,900
Daeen Chota*	Ghizer	PPAF-II	2004–06	15	64	576	Heterogeneous	39.9	Gravity	Spring	7,050
Diruch	Ghizer	KfW-I	2001–01	20	75	600	Homogeneous	42.0	Gravity	Spring	6,350
Dushkin*	Astore	PPAF-II	2004–06	15	203	1861	Homogeneous	32.8	Gravity	Spring	18,500
Duskhore Hashupi*	Skardu	PPAF-V	2007–08	13	144	1810	Homogeneous	40.7	Gravity	Nallah	3,000
Halpapa Astana	Skardu	EC	2010–12	9	77	714	Heterogeneous	29.4	Gravity	Spring	4,694
Hasis Paeen	Ghizer	KfW-I	1998–98	23	66	594	Homogeneous	30.9	Gravity	Spring	11,316
Hatoon Paeen*	Ghizer	KfW-II	2010–11	10	120	1080	Homogeneous	57.0	Gravity	Spring	19,200
Hundur Barkulti	Ghizer	7PV	2017–23	4	48	432	Homogeneous	46.7	Gravity	Spring	4,200
Hyderabad Center	Hunza	EC	2010–12	9	95	760	Homogeneous	31.0	Gravity	Spring	38,400
Janabad	Hunza	KfW-II	2014–14	7	75	626	Homogeneous	36.0	Gravity	Spring	14,500
Kirmin*	Hunza	PATRIP	2018–22	3	65	585	Homogeneous	55.7	Gravity	Spring	9,100
Kuno	Ghizer	KfW-II	2012–13	8	71	660	Homogeneous	61.9	Gravity	Spring	7,100
Marikhi	Ghizer	PPAF-VII	2011–12	9	36	288	Homogeneous	38.8	Gravity	Spring	3,100

(*Continued*)

Table 2.4 Continued

Settlement	*District*	*Funding*	*Period*	*Operation (years)*	*Total households*	*Total population*	*Community composition*	*Community share %*	*System type*	*Source*	*Pipe length m*
Nasir Abad Ishkoman	Ghizer	KfW-II	2010–11	10	53	477	Homogeneous	52.7	Gravity	Spring	11,360
Nazirabad*	Ghizer	KfW-II	2010–11	10	40	305	Homogeneous	24.7	Gravity	Spring	6,550
Rahimabad (Matumdass)*	Gilgit	JCVF	2014–14	9	120	896	Heterogeneous	44.4	Gravity	Spring	8,300
Shamaran Paeen*	Ghizer	PPAF-V	2007–08	13	24	288	Homogeneous	39.1	Gravity	Spring	4,700
Shilati*	Diamer	JCVF	2013–13	8	46	582	Homogeneous	52.9	Gravity	Spring	12,300
Singul Shyodass*	Ghizer	PPAF-I	2003–04	17	18	162	Homogeneous	26.3	Gravity	Spring	1,050
Staq Paeen	Skardu	KfW-I	2001–01	20	94	752	Homogeneous	30.0	Gravity	Spring	7,627
Sutopa	Skardu	PPAF-V	2007–08	13	83	1,017	Homogeneous	32.7	Gravity	Spring	4,400
Yuljuk	Skardu	KfW- I	2000–00	11	61	404	Homogeneous	42.5	Gravity	Spring	6,294

Note: *Engineering sub-sample

Table 2.5 WASEP urban sample overview

Settlement	*District*	*Funding*	*Period*	*Operation (years)*	*Total households*	*Total population*	*Community composition*	*Community share %*	*System type*	*Source*	*Pipe length m*
Aliabad Centre*	Hunza	KfW-II [IAUP]	2014–14	7	360	2,762	Homogeneous	29.0	Gravity	Nallah	22,500
Aminabad*	Gilgit	KfW-II	2012–13	8	300	2,800	Heterogeneous	16.4	Mechanized	Sump/ nallah	15,475
Amphary Patti (Mbad)	Gilgit	GoGB	2018–21	0	850	6,800	Heterogeneous	19.5	Gravity	Spring	47,961
Astore Colony*	Gilgit	GoGB	2016–18	3	210	1,680	Homogeneous	18.2	Mechanized	River	6,097
Chikas Kote	Gilgit	GoGB	2018–21	0	691	5,528	Heterogeneous	19.5	Gravity	Spring	58,262
Chokoporo Gahkuch Bala*	Ghizer	JCVF	2015–15	6	170	1,246	Homogeneous	34.6	Gravity	Spring	18,900
Diamer Colony*	Gilgit	GoGB	2016–18	3	288	2,304	Homogeneous	11.8	Mechanized	River	29,029
Domial Buridur Gahkuch Bala*	Ghizer	JCVF	2015–15	8	59	371	Homogeneous	42.1	Gravity	Spring	6,100
Hassan Abad Aliabad	Hunza	KfW-II	2010–11	10	77	693	Homogeneous	55.8	Gravity	Spring	12,087
Hunza Patti (Mbad)	Gilgit	GoGB	2018–21	0	551	4,408	Heterogeneous	19.5	Gravity	Spring	31,090
Hussainpura etc.	Gilgit	GoGB	2018–21	0	1,620	12,960	Heterogeneous	19.5	Gravity	Spring	112,000
Khanabad*	Ghizer	JCVF	2015–15	6	55	421	Homogeneous	47.7	Gravity	Spring	6,100
Noor Colony Urban Project*	Gilgit	KfW-II	2011–12	9	325	3,541	Heterogeneous	13.5	Mechanized	River	7,882

(Continued)

Table 2.5 Continued

Settlement	*District*	*Funding*	*Period*	*Operation (years)*	*Total households*	*Total population*	*Community composition*	*Community share %*	*System type*	*Source*	*Pipe length m*
Noorabad Extension*	Gilgit	KfW-II	2014–14	7	325	2,600	Homogeneous	17.8	Mechanized	River	18,590
Princeabad Bala etc.	Gilgit	GoGB	2018–21	0	992	7,936	Heterogeneous	19.5	Gravity	Spring	55,974
Rahimabad Aliabad*	Hunza	KfW-II [IAUP]	2011–12	9	321	2,204	Homogeneous	26.9	Gravity	Nallah	21,400
Sakarkoi	Gilgit	KfW-II	2011–12	9	116	1,047	Heterogeneous	55.8	Mechanized	River	7,882
Shangote Patti (Mbad)	Gilgit	GoGB	2018–21	0	590	4,720	Heterogeneous	19.5	Gravity	Spring	33,291
Sharote etc.	Gilgit	GoGB	2018–21	0	630	5,040	Heterogeneous	19.5	Gravity	Spring	35,548
Soni Kot*	Gilgit	KfW-II	2010–11	10	310	2,790	Heterogeneous	59.0	Mechanized	River	16,371
Sultanabad 1&2	Gilgit	GoGB	2018–21	0	799	6,392	Heterogeneous	19.5	Gravity	Spring	45,084
Syedabad	Gilgit	GoGB	2018–21	0	292	2,336	Heterogeneous	19.5	Gravity	Spring	16,476
Wahdat Colony*	Gilgit	GoGB	2016–18	3	372	2,976	Homogeneous	11.6	Mechanized	River	17,730
Yasin Colony*	Gilgit	GoGB	2016–18	3	170	1,360	Homogeneous	11.8	Mechanized	River	13,833
Zulfiqarabad*	Gilgit	GoGB	2016–18	3	469	3,752	Homogeneous	11.8	Mechanized	River	25,160

Note: *Engineering sub-sample

Table 2.6 Gilgit City projects: Jutial

Actual WASEP projects	*Funding*	*WASEP sample*	*Households*	*Population*	*Community composition*	*Settler status (majority)*
Astore Colony	GoGB	Astore Colony	61	488	Homogeneous	New settlers
Azam Block Prince Colony	GoGB					
Diamer Colony	GoGB	Diamer Colony	288	2,304	Homogeneous	New settlers
Hussainabad Colony	GoGB					
Main Zulfiqarabad	GoGB					
Nawaz Sharif Colony	GoGB					
Noor Colony Urban Project	KfW-II	Noor Colony Urban Project	325	3,541	Heterogeneous	Old settlers
Noorabad Extension	KfW-II	Noorabad Extension	325	2,600	Homogeneous	New settlers
Satellite Town Colony	GoGB					
Shami Muhala	GoGB					
Wahdat Colony	GoGB	Wahdat Colony	372	2,976	Homogeneous	New settlers
Yasin Colony	GoGB	Yasin Colony	170	1,360	Homogeneous	New settlers
Zulfiqarabad	GoGB	Zulfiqarabad	469	3,752	Homogeneous	Old settlers

Table 2.7 Gilgit City projects: Danyore

Actual WASEP projects (mohallah)	*Funding*	*WASEP sample*	*Households*	*Population*	*Community composition*	*Settler status (majority)*
Amphary Patti (Muhammadabad)	GoGB	Amphary Patti (Mbad)	850	6,800	Heterogeneous	Old and new settlers
Chikas Kote: Chikas Kote, Baig Market	GoGB	Chikas Kote	691	5,528	Heterogeneous	Old settlers
Hunza Patti (Muhammadabad)	GoGB	Hunza Patti (Mbad)	551	4,408	Heterogeneous	Old and new settlers
Hussainpura etc.: Magrote, Hussainpura, Mayon Gali, Ziarat Mohallah, Qasimabad 1&2, Majukal, Shamsabad, School Area	GoGB	Hussainpura etc.	1,620	12,960	Heterogeneous	Old settlers
Nizarabad etc.	GoGB					
Princeabad Bala etc.: Princeabad Bala, Chandni Chowk Area, Sikarkui, Butary Het (New Hussainabad)	GoGB	Princeabad Bala etc.	992	7,936	Heterogeneous	Old settlers
Princeabad Paeen, Shangote, Medhiabad Colony, Nagar Muhalla, Sharote	GoGB					
Shangote Patti (Muhammadabad)	GoGB	Shangote Patti (Mbad)	590	4,720	Heterogeneous	Old and new settlers
Sharote etc.: Sharote, Shabki Hit, Saito Hit, Akhori Hit, Thokihit	GoGB	Sharote etc.	630	5,040	Heterogeneous	Old settlers
Shereenabad etc.: Shereenabad, Malki Hit, Baki Hit, Puri Hit, Astan Muhalla, Magrote & Sajidabad Maungali	GoGB					
Sultanabad 1 & 2	GoGB	Sultanabad 1 & 2	799	6,392	Heterogeneous	Old and new settlers
Sultanabad 3	GoGB					
Sultanabad Golodass	GoGB					
Syedabad	GoGB	Syedabad	292	2,336	Heterogeneous	Old and new settlers

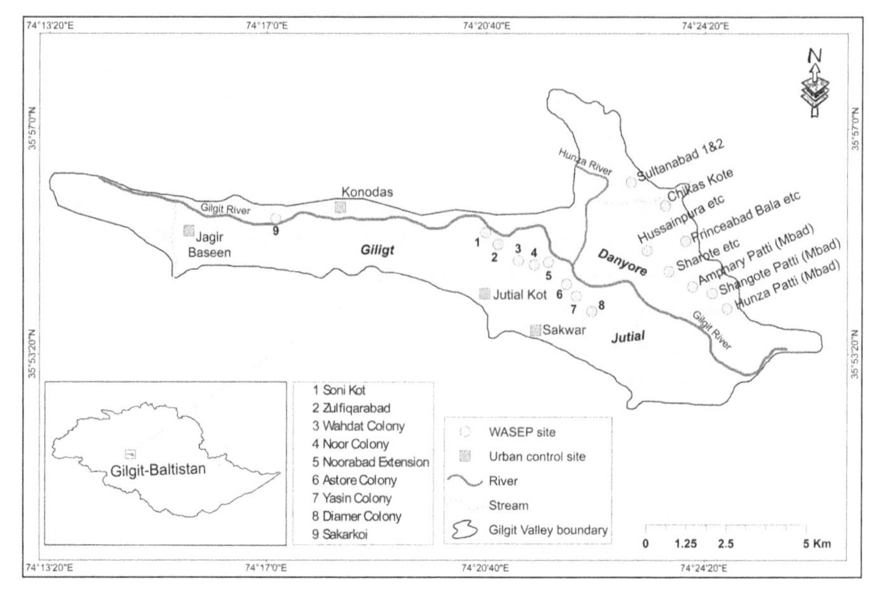

Figure 2.3 WASEP urban sites, Gilgit City
Source: map by Karamat Ali and Garee Khan

Jutial and Danyore thus represent a significant scaling up of the WASEP model, being notably larger and catering to a more mixed population (particularly in Danyore) than previous (rural) projects and are thus of special interest for the purposes of this research. Both umbrella schemes also illustrate the impact of conflict and the engineering features of urban projects. Figure 2.3 locates the Jutial and Danyore sample projects on either side of Gilgit River that runs through Gilgit Valley, where a major difference is access to clean water sources.

The lack of water rights and hence access to upstream water for Jutial residents necessitated a mechanized water system to extract water from Gilgit River and provide water treatment via riverbank filtration (Photo 2.1). Mechanized systems such as this are not only more expensive to construct but also have higher running costs unlike the gravity-fed system in Danyore and rural sites. As with Jutial, Danyore projects all share a common water source that is then allocated to the different neighbourhoods/WSCs through a distribution chamber and based on agreed water shares for each *mohallah* (see Photo 2.2).

From the two groups of 25 rural and 25 urban projects, a sub-sample of 12 rural and 14 urban projects was generated for engineering audits (marked with an asterisk in Tables 2.4 and 2.5). Data was provided for eight engineering variables: water source, system type, tanks, treatment, water quality, length of pipe, distance from source, and environmental vulnerability. Four engineering variables displaying unrelated variation were used to structure the rural engineering sample: source, water treatment, length of pipe used, and distance from water source (as a proxy for vulnerability to natural hazards). Only one

Photo 2.1 Jutial: Pumping station with riverbank filtration
Photo credit: Jeff Tan

rural sample project used stream (*nallah*) water and only one used water treatment. Both were selected for inclusion. Projects were then divided into a matrix by length of pipe used and distance from water source. Five projects were sampled from this matrix proportionally to the distribution of these variables in the rural sample (as engineering data was only available for the sample of 25 rural). The remaining six rural engineering audit sample projects were randomly sampled, one from each district group (Table 2.4).

For the urban engineering sample, urban projects were first stratified by district – one project from the Ghizer group and one from the Hunza-Nagar group were sampled to include the variation in water quality and cost. For Gilgit urban projects, engineering variables all show strong relationships with water source. (The engineering data and results are discussed in Chapters 9 and 10.) To capture socio-spatial variation, projects were quota sampled by area, then water source, then by the majority group if known (e.g. Shia, Sunni, Ismaili, Norbakshi) as follows: three from Danyore (one from each majority group); four from Upper Jutial (one from each majority group); two from Lower Jutial (one from each majority group); and two each from Sakarkoi and Soni Kot. Six rural control sites without WASEP projects (one each from Astore, Diamer, Ghizer, Gilgit, Hunza, and Skardu) were chosen on the basis of their similarity to WASEP projects in the study. Six urban control sites were chosen for their similarity to WASEP projects in the study, one from Ghizer, one from Hunza-Nagar, and four from Gilgit City.

Households were included in the study through both telephone and door-to-door interviews. The number of respondents required was calculated (based on total household and population numbers provided by AKAH) to meet the minimum required for valid random sampling using the Slovin formula (to calculate the sample size given the population size and a margin of error)

(a)

(b)

Photo 2.2 Danyore: (a) Water distribution chamber and (b) water allocation, under construction
Photo credit: Jeff Tan

(see 'household responses' in Table 2.2). For the telephone survey, a protocol was provided to systematically conduct random door-to-door phone number collection where phone numbers were first collected in the sample area and then randomly selected from this larger pool to be contacted. In the door-to-door survey, households were selected for inclusion using systematic random sampling to select every second to tenth house (depending on the size of the settlement).

The 3,132 usable responses from the household survey (Table 2.2) were supplemented by interviews with all WSC treasurers, 20 rural and 20 urban village or local social mobilizers, and all available women water and sanitation implementers who are responsible for supporting health and hygiene education (see Chapter 6). Based on a preliminary analysis of the household survey data, focus group discussions were conducted in 14 rural and 12 urban projects where specific issues were identified for follow up. The following chapters present the synthesis and analysis of these findings.

2.5 Conclusion

This chapter has introduced the data and some of the evidence, as well as highlighted some of the key features of CBWM to be analysed in the chapters that follow. These features centre on the differences between rural and urban projects and the impact of community diversity and unity on participation levels in WASEP projects (Chapter 5), conflicts that occurred (Chapter 7), tariff payment and hence financial sustainability (Chapter 8), and finally responses to natural hazards (Chapter 10). Differences are also reflected in the engineering challenges in more complex and expensive urban water systems (Chapters 8, 9, and 10). It was necessary to prioritize urban projects in this study, and specifically the umbrella schemes in Jutial and Danyore, in order to better understand the differences between rural and urban WASEP DWSS, and the difficulties faced in urban projects.

In particular, the Jutial and Danyore schemes highlight the issues of water access, conflict, and engineering solutions. In the case of Jutial, the absence of traditional water rights had engineering and financial implications with the need for a mechanized system and recurring electricity costs for pumps. For Danyore, the number of projects and *mohallah* involved increased the likelihood of conflict over the allocation of water between projects/neighbourhoods as well as (political) opposition to WASEP. As a result, the Danyore project, due for completion in 2021, was significantly delayed and had not been completed in the course of the fieldwork for this research. These challenges of scaling up make WASEP urban projects in Jutial and Danyore especially instructive.

CHAPTER 3

Water governance and related institutions in Gilgit-Baltistan

Jeff Tan

3.1 Introduction

External support for communities is central for the sustainability of community-based water management (CBWM). While support is expected from government, donors, and NGOs, government plays the largest role. Wider governance structures are needed at all levels of service delivery for a system-wide approach through the creation and alignment of legislation, policies, and institutions (World Bank 2017). The effectiveness of external support for CBWM thus depends on the capacity of the state and quality of related institutions. However, low-income countries are characterized by weak state capacities and fragmented institutions, and hence the ineffective governance of water supply and sanitation (WSS) services (World Bank 2017). In the case of the Water and Sanitation Extension Programme (WASEP), effective state support is largely constrained by weak state capacity and institutional fragmentation on a federal level that is multiplied at the provincial level and compounded in Gilgit-Baltistan because of its unresolved constitutional status. This significantly weakens the ability of the Government of Gilgit-Baltistan (GoGB) and state institutions to provide long-term support to WASEP communities and CBWM, and highlights the institutional and governance challenges that constrain the delivery of sustainable WSS services in Gilgit-Baltistan.

This chapter provides the wider institutional context that frames how WASEP delivers water services in Gilgit-Baltistan, and how this affects external support from government and the state. The next section provides an overview of water governance at the federal and provincial levels. Section 3.3 looks at how institutional fragmentation and weak state capacity is reflected in legislation, policies, institutions, monitoring and regulation, and human resources, drawing from official reports. Section 3.4 examines how these institutional and capacity constraints are mirrored in Gilgit-Baltistan. Section 3.5 briefly outlines how these institutional weaknesses and fragmentation have impacted external support for, and hence the sustainability of, WASEP.

3.2 Water governance in Pakistan

Weak state capacity and institutional fragmentation in Gilgit-Baltistan can be viewed as part of wider problems in water, sanitation, and hygiene (WASH) governance in Pakistan that are in turn a reflection of underlying public administration inefficiencies related to the legacy of colonial-era bureaucratic structures (Muhula 2019) and the complexities of federal systems of government (Young et al. 2019). Weak state capacity is in part due to the absence of strong human resource management systems, including basic staffing data and monitoring systems, making it difficult to estimate staff strength and plan for recruitment at the appropriate levels and skillsets (Muhula 2019). Institutional fragmentation is a common feature of federal systems for water governance that are often 'a complex patchwork of institutions, policies, and legal provisions at provincial and national levels' (Young et al. 2019: 54).

Institutional arrangements for WASH in Pakistan can be traced back to the 1973 Constitution which assigned WSS services to provincial governments but where 'relevant policies, institutions, and legal provisions are distributed across the national and provincial levels' and national institutions 'coexist with, and sometimes overlap with, provincial institutions, and the legal framework for each province includes its own laws and regulations overlain by relevant national provisions' (Young et al. 2019: 54). The 2001 Local Government Ordinances further transferred political, administrative, and fiscal powers from higher to lower tiers of government 'in order to bring governments closer to common citizens for greater accountability and better understanding of the needs and preferences of people' (Anjum 2001: 845).

The 18th Amendment to the Constitution in 2010 strengthened the role of provincial governments in WSS with each developing its own legislation, policies, and plans to provide services (Young et al. 2019: 62). There was thus no specific ministry in charge of WSS at the national level, and limited involvement of national institutions (Cooper 2018). The result is the involvement of multiple federal ministries in WASH, including the Ministry of Climate Change (MoCC), Ministry of Health Service Regulation and Coordination, Ministry of Federal Education and Professional Training, Ministry of Agriculture, and the Planning Commission of the Ministry of Planning, Development and Reforms. MoCC (formerly the Ministry of Environment before it was devolved) assumed WASH leadership in 2011 at the federal level for policy formulation, standards settings, reporting, and coordination (GoP 2019a).

WASH policies included the National Environmental Policy 2005, the National Sanitation Policy 2006, the National Drinking Water Policy (NDWP) 2009, the National Behavioural Communication Strategy, and National Climate Change Policy 2012 (UN-Water 2014). The NDWP was drafted in 2003 by the Ministry of Environment to improve 'water access, treatment, and conservation through enhanced community participation and public awareness, cost-effective infrastructure, research and development, and PPPs [public–private partnerships]' (Young et al. 2019: 62). The MoCC is still revising and aligning the NDWP with the United Nations Sustainable Development Goals (SDGs) and Pakistan's development agenda to provide national guidelines for provincial governments to prepare their own policies and strategies 'as per local resources, climate and situation' (GoP 2019a: 10–11; World Bank 2016).

Provincial governments 'have retained the responsibility for policy provision, human resource management and financing' with Local Government Ordinances (2001, 2011, 2013) transferring WSS services to newly created local government institutions (World Bank 2016: 10). Provincial drinking water policies provide the framework to guide and support district governments, local government institutions, water utilities, and communities for improving WSS services although provincial institutions 'have largely resisted reforms because of entrenched and contested interests, amplified by a lack of capacity' (Young et al. 2019: 62). Four provinces (Punjab, Khyber Pakhtunkhwa, Sindh, and Balochistan) approved provincial drinking water policies as of 2011, with the semi-autonomous provinces of Azad Jammu Kashmir and Gilgit-Baltistan still in the process of revision and approval (GoP 2019a: 10–11).

Provincial WASH services are delivered through implementing agencies on a district, sub-district (*tehsil*), and village level. On the district level, Public Health Engineering Departments (PHEDs) provide rural drinking water and sewerage infrastructure for populations above 500 people with Local Government and Rural Development Departments (LG&RDDs) responsible for rural populations under 500 people. Provincial Health Departments are responsible for hygiene, and Education Departments cover WASH services in schools (Lerebours 2017, cited in Cooper 2018). On the sub-district (*tehsil*) level, Tehsil Municipal Administrations provide WSS services as part of municipal services, and Development Authorities and Water and Sanitation Agencies (WASAs) deliver WSS services to large cities, although PHEDs often end up undertaking construction because of their better engineering capacities (Lerebours 2017, cited in Cooper 2018). Finally, on the village level, union councils lead projects, and employ Lady Health Workers and Sanitary Officers for health and hygiene education (Lerebours 2017, cited in Cooper 2018).

3.3 Institutional fragmentation and weak state capacity

Water governance in Pakistan has been constrained by institutional fragmentation and weak state capacity. Federal and provincial government reports provide clear evidence of the ways in which these constraints impact WASH delivery. The UN-Water Global Annual Assessment of Sanitation and Drinking-Water (GLAAS), and WASH Joint Sector Reviews (JSRs – a collaboration between MoCC and UNICEF since 2016 and extended to four provinces) identify key bottlenecks, provide plans and resources, and allocate responsibilities to improve governance (GoP 2019a). Institutional fragmentation and weak capacity are reflected in legislation, policies, institutions, monitoring and regulation, and human resources.

The lack of exclusive legislation at the federal and provincial levels has undermined effective water governance and management (Lerebours 2017, cited in Cooper 2018). Fragmented, overlapping, and contradictory WASH legislation include the Provincial Local Government Act 2013, provincial Environment Acts, and draft Municipal Acts. The Local Government Act 2013 assigns WSS service provision, access, and operations and maintenance (O&M) to local councils in urban and rural areas but O&M responsibility in rural areas is entrusted to local communities through community-based organizations (GoP 2019a).

Provincial Environmental Protection Acts mandate Environmental Protection Agencies (EPAs) to monitor Environment Quality Standards but without comprehensive legislation that empowers EPAs to prepare regulatory rules and standards. Compounding these challenges, the national standards for service quality are not legally binding, rendering enforcement distinctly problematic. Policy frameworks 'do not adequately separate institutional roles for water supply, asset ownership or management, and service delivery, and the absence of an independent regulator further undermines progress' (Young et al. 2019: 62). The NDWP (2009) has yet to be implemented, and only two provincial WSS policies are aligned with the SDGs.

Fragmentation is reflected in overlapping departments, and duplication of roles and responsibilities across multiple institutional layers. Further exacerbating the situation are ambiguities in the authority of key departments, unclear reporting lines, and coordination at most levels on a needs basis, with numerous actors working for a common goal but pulling in different directions at the national, provincial, and local level and between sectors (domestic, agriculture, industrial) (Lerebours and Villeminot 2017; State Bank of Pakistan 2017; GoP 2019a). Service delivery 'is spread across many institutions with varying capacities, differing reporting lines, and limited coordination' that include PHEDs, local government departments (LGDs), and WASAs, with PHEDs and WASAs delivering urban services in large cities despite the fact that LGDs have broad service delivery responsibility (Young et al. 2019). LGDs are responsible for water services but provincial departments such as PHED and LG&RDD continue to play a leading role in resource allocation, project identification, service delivery, and O&M.

Monitoring and regulation are constrained by an inadequate national information system and incomplete data, particularly for rural planning, with data only used for a minority of decisions (see UN-Water 2014). The existence of multiple departments and institutions makes it difficult to obtain comprehensive data to track WASH coverage. This significantly impacts the efficiency and effectiveness at provincial and federal levels for identification, implementation, and monitoring. Performance indicators to track functionality are undeveloped and outcome data is not collected. As a result, there is no clearly defined performance base or cross departmental guidelines and standards for planning and evaluation of WASH schemes. There is thus little to no testing of water quality against national standards and no regulatory authority for setting tariffs or service quality.

Weak state capacity is related to human resource constraints, last estimated at 50% of what is required for data collection and monitoring (IUCN 2014, cited in Cooper 2018). There are insufficient human resources, particularly skilled graduates and workers, for planning, design, implementation, monitoring, and reporting, with no systematic capacity development or defined capability framework to develop technical and management capacities. Local council representatives in particular lack capacity and a clear understanding of their role. This is compounded by a lack of continuity in policy and personnel, with the inevitable change of leadership and policy on the provincial level that follows any change of political leadership on the federal level, and regular transfers of staff at the departmental level.

3.4 Water governance in Gilgit-Baltistan

Water governance in Gilgit-Baltistan is complicated by its unresolved constitutional status, partial democratic representation, and complete fiscal dependence on the federal government (Ahmed et al 2010). Previously known as the Northern Areas (along with Azad Jammu and Kashmir), this semi-autonomous province remains constitutionally undefined following the 1948 and 1949 United Nations resolutions 'attesting to the status of present-day Gilgit-Baltistan as a disputed territory waiting for a plebiscite to decide the fate of Kashmir' (Holden 2019: 3). It is administered by the federal government through the Gilgit-Baltistan Legislative Assembly (GBLA), subsequently renamed Gilgit-Baltistan Assembly (GBA), the provincial apex body of locally elected representatives. The absence of full democratic representation at the national level and weak administrative capacity limits state capacity; GBLA's legislative and administrative authority is historically very limited in comparison with other provinces (Ahmed et al 2010). While the GBLA is 'empowered to pass the annual budget and Annual Development Program (ADP), it must be approved by the Governor, and virtually all resources come from the federal government' (Ahmed et al 2010: 23).

The Gilgit-Baltistan Local Government Act 2014 established 'an elected local government system to devolve political, administrative and financial responsibility and effective delivery of services through institutionalized participation of the people at local level' (GBLA 2014: 1). Water quality and service delivery is covered by different legislations. The Environmental Protection Act of Gilgit-Baltistan 2014 sought 'to provide for the protection, conservation, rehabilitation and improvement of the environment, prevention and control of pollution, and promotion of sustainable development' (GoGB 2015). The Gilgit-Baltistan Empowerment and Self-Governance Order 2009 (ESGO) provided the basis for the 'necessary legislative, executive and judicial reforms for granting self-governance' (GoP 2019a).

The Development of Cities Act 2012 was 'to establish a comprehensive system of planning and development in order to improve the quality of life in the cities ... relating to the improvement at [sic] the ... water supply, sewerage, drainage, solid waste disposal and matters connected therewith' (GoP 2020: 689). The delivery of WSS services is meant to be framed by the Draft Gilgit-Baltistan Drinking Water Policy 2019 that provides 'a broad framework and guidelines for improving service coverage of safely managed drinking water' (GoGB 2019: 4) to accelerate and provide universal coverage by 2030 (GoP 2019b). It was prepared and submitted by the WASH Unit of the LGD for approval by GBLA and was endorsed by the chairs of Standing Committees of GBLA on 18 June 2019. However, it was still awaiting approval from the Cabinet of Ministers and GoGB in 2023.

The Draft Drinking Water Policy lists 11 principles:

1. Equitable access and government responsibility
2. Water allocation for drinking
3. Integrated drinking water service with health, education, and nutrition
4. Women's participation
5. Multi-sectoral approach
6. Delegation of responsibility
7. Financial and institutional sustainability through community and private sector involvement, and environmentally appropriate, climate change resilient, and low-cost technologies
8. Mitigation of future risks from water demand and climate change
9. Review of ownership and traditional water rights, and enactment of laws for equitable and safe drinking water
10. Financial allocations to enhance coverage and improve water availability, quality, and O&M
11. Streamline inter-departmental roles and responsibilities.

From these largely aspirational principles, a number of policy guidelines are put forward to:

- increase access and coverage; protect and conserve water resources; introduce legislation;
- provide water treatment for safe drinking water;
- develop appropriate technologies and standardization;
- encourage community participation;
- build capacity;
- explore and support public–private partnerships (PPPs);
- undertake research and development for appropriate technologies;
- structure institutional arrangements for implementation and coordination; and
- increase preparedness for disaster risk reduction.

Provincial water-related institutions are largely replicated in Gilgit-Baltistan. At the district level, City Metropolitan Corporations cover the urban centres of Gilgit and Skardu with populations over 100,000 and 'provide, operate, manage, and improve the municipal infrastructure and services', including water supply, and the control and development of water resources (GBLA 2014: 28). At the sub-district (*tehsil*) level, municipal corporations (covering urban populations above 50,000), municipal committees (urban populations over 30,000), and town committees (urban populations above 10,000) 'provide, operate, manage, and improve the municipal infrastructure and services' including water supply, control, and development of water resources (GBLA 2014: 28). At the village (rural) level, union councils manage and provide rural water supply schemes and public sources of drinking water as part of local municipal services. Tehsil councils assist union councils in the 'provision and maintenance of rural water supply schemes and public sources of drinking water, including wells, water pumps, tanks, ponds and other works for the supply of water'; and district councils provide WSS services for rural areas (GBLA 2014: 25).

This overview of WSS legislation, policies, and administration hints at some of the issues of water governance in Gilgit-Baltistan. The policy guidelines for the Draft Drinking Water Policy serve to illustrate the current institutional and governance weaknesses and gaps identified on the national level in Section 3.3. These include: inadequate legislation, standards, (inter-sectoral) policies, strategies, and action plans; institutional fragmentation and duplication; lack of integration, (inter-sectoral) coordination, oversight, and policy implementation; weak state capacities (including human resources) in particular for water quality testing, monitoring and regulation; and crucially, the lack of rehabilitation, upgradation, O&M and hence sustainability due to insufficient user charges, cost recovery, and cross subsidization.

Specific evidence of institutional fragmentation and weak capacity can be gathered from federal and provincial government reports, GLAAS (UN-Water 2014), and JSRs (GoP 2019a, 2019b) covering legislation, policies, institutions, monitoring and regulation, and human resources. Official reports identify considerable overlap between the laws, frameworks, and regulations, which creates ambiguities for different stakeholders with the existing dichotomy between regulations resulting in unnecessary delays (GoP 2019b: 14). Examples of this includes ESGO 2009 that does not define government roles and responsibilities, and the Local Government Act 2014 that assigns union councils responsibility for rural water services and public sources of drinking water, but with substantial funds allocated to the Public Works Department (PWD), and the absence of a well-defined accountability mechanism in the legal framework (GoP 2019b). Provincial water policy centres on the 2019 Draft Drinking Water Policy that identifies problems and makes recommendations, but has still not been approved by GBLA, with complaints about the lack of a federal government plan or 'Master Plan' to guide provincial governments and the absence of a tangible commitment for WASH allocations (see e.g. GoP 2019b).

Institutional fragmentation is mirrored at the provincial level in Gilgit-Baltistan where 'the situation in relation to [the] roles of stakeholders including public sector institutions is very complex for [the] water and sanitation sector' (GoP 2019b: 15). WSS services are provided by a number of overlapping departments with initiatives 'taken in silos' resulting in poor information sharing, a lack of consultation and deliberation among government departments and other stakeholders, friction between departments, and the evasion of responsibility (GoP 2019b: 15). While LG&RDD is the designated lead institution for WASH-related interventions, 'the policy directives and business rules provide space for participation of other departments in WASH related interventions', leading to fragmentation and overlapping roles and responsibilities of service provider institutions (GoP 2019b: 15). PWD is responsible for the supply of drinking water and wastewater management in Gilgit-Baltistan, with the drinking water supply component under the responsibility of the Secretary of Works at the provincial level and the Superintendent Engineer at the district level.

There is also 'very little coordination between the relevant institutions that are concerned with [the] supply of drinking water and its quality assurance in urban settlements of GB' (GB-EPA 2012: 16). As previously noted, the Local Government Act 2014 delegates drinking water and sanitation provision to local government, but PWD is primarily responsible for planning, design, construction, and O&M of development projects (GoP 2019b). GB-EPA conducts water quality tests of sources for installation of water supply schemes but lacks resources and capacity for a comprehensive water quality surveillance programme in the province. However, the WASH Unit in LG&RDD also tests water quality for all drinking water schemes in Gilgit-Baltistan by all departments but only focuses on 'rural areas and lacks pre-emptive strategies either to control pollution or maintain the source water quality in the networks' (GB-EPA 2012: 16). To confuse matters further, water quality monitoring in Pakistan is meant to be the sole responsibility of the Pakistan Council of Research in Water Resources, with the federal government itself seemingly unclear about which department should be responsible for testing water quality.

Monitoring and regulation are undermined by institutional fragmentation and the lack of data which makes it difficult to set standards, performance targets and budgets, and to plan. Databases are fragmented with individual departments maintaining their own datasets with no consolidation of departmental reports. There is no data on overseas development assistance and no system to track 'significant investment by donor [sic] for NGOs [sic] programme related to WASH' as this is not reported or tracked at the provincial level (GoP 2019b: 42). All funding in Gilgit-Baltistan is allocated under the ADP but there is no published data for budget allocations and expenditures, and the WASH budget is aggregated with no separate codes for sub-sectors, making the tracking of provincial WASH allocation and expenditure 'difficult and even impossible' (GoP 2019b: 41).

Without data, there are no performance measures or monitoring of service delivery (in terms of functionality, hours of service, affordability, quality, quantity, and cost effectiveness) and service and coverage targets are unrealistic given the available resources and capacity. As a result, there are no provincial plans with clear targets, activities, indicators, timelines, and budgets; no monitoring and surveillance framework for regular monitoring of established standards; and no regulator or institution with clear regulatory functions (GoP 2019b). These weaknesses are in part related to the wider problem of insufficient human resources previously noted, with institutions lacking 'capacity to fulfil sector roles and responsibilities' for sustainable service delivery at scale, including 'the availability of necessary structures, tools, training, and incentives' (GoP 2019b: 46). There is no government-led capacity development plan, training needs assessment, or assessment for human resource strategy because of insufficient resources for capacity building, with the PKR400 m (US$3.5 m) allocated to WASH under the ADP in 2017–18 'not good enough to fulfil all sector requirements' (GoP 2019b: 39).

3.5 State institutional constraints and WASEP

Weak state capacity and institutional fragmentation in Gilgit-Baltistan affect the ability of the state to provide long-term support to WASEP communities as recommended in the CBWM literature and official reports. This is reflected in legislation, policies, institutions, monitoring and regulation, and human resources. Legislation is necessary to promote PPP and support community management. The Draft Drinking Water Policy 'policy guideline' includes exploring the possibility of PPPs for urban areas, creating an enabling environment 'for private sector and non-governmental organizations to invest and support government efforts', and 'to replicate the model of Water and Sanitation Extension Program (WASEP) and successful government partnerships in other sectors under PPP' (GoGB 2019: 10).

PPP thus provides the legal basis for GoGB's partnership with the Aga Khan Agency for Habitat (AKAH) and hence the implementation of WASEP. However, there is no framework for PPPs in Gilgit-Baltistan and no PPP legislation was in place when the memorandum of understanding for the Danyore scheme was signed by GoGB, AKAH, and the community. The PPP Act was subsequently adopted and passed in August 2019, and assented in November 2019, but this was taken from similar legislation in Sindh province, with the issue being that there is no substantial 'private sector' to partner in Gilgit-Baltistan and no provision for agreements between government and communities. The legal status of the WASEP model in relation to the PPP Act thus remains unclear.

It also remains to be seen if legislation to clearly define community ownership and authority, and to support the enforcement of tariff collection and sanctions for non-payment, is likely given the problems of poorly framed current legislation and the inability of GBLA to approve policies, most notably the Draft Drinking Water Policy. Recommendations in the literature for legal ownership by communities of water assets are problematic given the historical issue of water rights and access needed for piped networks (see Chapter 7). This lack of (legal) authority is a common theme across the literature and is echoed in concerns raised by AKAH and some water and sanitation committees (WSCs) about the absence of a legal framework to formalize the official status of WSCs that would allow registration with local authorities and hence access government funds (allocated to public water schemes) and bank loans, and to be authorized to collect tariffs and enforce (graduated) sanctions. This however assumes that the state has the additional capacity and resources to support WSCs in these actions (e.g. through policing and the court system) at a time when financial resources are limited and human capacities lacking. An example is when GoGB did not step in to resolve conflict over the allocation of water that delayed the implementation of the largest WASEP project in Danyore (see Chapter 7).

The impact on communities of direct state interventions through enforcement and sanctions in the case of tariff payments also raises fundamental questions about community solidarity and unity in the CBWM model given the wider social ramifications of a legalistic approach (e.g. by destabilizing relationships with relatives, friends, and neighbours) (see e.g. Broek and Brown 2015; Chowns 2019). Recommendations in the JSR to '[p]ilot the concept of tariff initially for O&M services of Water and Sanitation through legislation' (GoP 2019b: 22) is an acknowledgement of the anticipated resistance to introducing and legislating tariffs, and a reflection of both the inherent problem of irregular and non-payment of tariffs in general and low levels of state capacity.

The number of overlapping departments and evasion of responsibility (GoP 2019b: 15) has made it difficult for WSCs that require external (financial) support for major repairs. This is not helped by the legal ambiguity of departmental responsibility once WASEP projects are completed and handed over to communities. While AKAH legally ends any liability and obligation after the project handover, it is far from clear who in fact legally owns the water infrastructure and is hence responsible for major repairs and subsequent refurbishments and service expansion. As communities do not have legal ownership of water assets, and given that WASEP projects are often implemented on behalf of, if not funded by, GoGB, it would appear that the state ultimately retains ownership and responsibility. However, without long-term financing beyond the delivery of WASEP schemes, and institutional fragmentation, there is little incentive for any of the existing provincial state institutions to assume responsibility for maintenance. At the same time, without monitoring, including regular water quality testing, AKAH does not know if water in WASEP projects is still safe or if water systems are functional or require repairs or rehabilitation.

3.6 Conclusion

The evidence from official reports, and acknowledgement by government, illustrates the scale and depth of the problems brought about by institutional fragmentation and weak state capacities and how these invariably constrain the ability of government and the state to provide the necessary external support to sustain WASEP. The challenge involves both the fiscal resources needed to fund capacity building, for example through training and recruitment, as well as knowing where to start, given the lack of data and records. Most critically perhaps is that sustainable service delivery institutions must have the capacity to identify underlying problems and implement appropriate solutions.

These challenges are constrained by existing structures, hierarchies, and inter-departmental rivalries that have been compounded by Pakistan's federalism and decentralization policies, and may be intractable without top-down reform of state institutions and external intervention. Interventions have been made by UN-Water, UNICEF, and the World Bank through GLAAS and JSRs. However, these interventions do not appear to recognize the inherent limitations of the CBWM model, and are guided by an ongoing faith in the private sector as illustrated in the continued promotion of PPPs (where there may not be a significant private sector) rather than focusing on rebuilding state capacities.

CHAPTER 4

The WASEP model of community management

Anna Grieser, Saleem Khan, and Jeff Tan

4.1 Introduction

The Water and Sanitation Extension Programme (WASEP) represents a near textbook implementation of the community-based water management (CBWM) model, applying the principles of participation, ownership, control, and cost sharing. However, WASEP is also distinct in incorporating health and hygiene (H&H) awareness, engineering solutions, and water quality management as part of its integrated approach. Additionally, WASEP schemes are very much initiated and led by the implementing NGO, the Aga Khan Agency for Habitat (AKAH), contrary to the demand- or community-led approach of CBWM. This chapter provides an overview of the key features of WASEP to frame the discussions in the following chapters.

WASEP was introduced and implemented by the Aga Khan Planning and Building Service (AKPBS), an agency of the wider Aga Khan Development Network (AKDN), in 1997. This followed from the 'Water, Sanitation, Hygiene, and Health Study Project' (WSHHSP) (1993–1996), a three-year study by Aga Khan Health Services (AKHS) into the failure of previous government and AKDN drinking water supply schemes (DWSS) (AKHS 1997). WSHHSP sought to understand why earlier community managed water projects failed. Out of 862 villages examined, 502 (58%) had water supply schemes of which 85 (17%) were non-functional, 32 (6%) were unfinished, 127 (25%) had incomplete coverage, 172 (34%) required major repairs, and only 86 (17%) were possibly satisfactory, not accounting for uninterrupted supply and water quality (Ahmed et al. 1996; Ahmed and Alibhai 2000; Hussain et al. 2000). More critically, the study found exceedingly high levels of *E. coli* contamination, with waterborne diseases accounting for 50% of all infant deaths in Gilgit-Baltistan and neighbouring Chitral, and households spending on average PKR1,000 (around US$25)[1] annually on treating diarrhoea.

WSHHSP recognized that engineering solutions alone could not deliver health benefits. Its findings and recommendations provided the impetus for WASEP's integrated approach that focused on: (i) engineering solutions to deliver clean drinking water and sanitation, particularly for unserved areas; (ii) capacity building of community organizations and state agencies for the management and support of water systems; and (iii) H&H education to bring about behavioural change for improved health outcomes (see Table 4.1).

This integrated approach is guided by CBWM principles of community participation, ownership, and cost recovery, and centred on social mobilization and community-based financing. Social mobilization is the cornerstone of WASEP's participatory approach and key to mobilizing communities to demonstrate demand for water services, build capacities, resolve conflicts, promote H&H awareness and behavioural change, and establish community-based financing to promote community self-reliance for the maintenance of water systems. These are reflected in the key features of WASEP that are discussed below: participation (Section 4.2); ownership and control (Section 4.3); cost sharing (Section 4.4); H&H promotion (Section 4.5); engineering solutions (Section 4.6); and water quality management (Section 4.7).

4.2 Participation

Participation in WASEP needs to be understood in terms of the wider social context and history of community organizations, committees, and community water management in Gilgit-Baltistan. Community-based water supply projects were introduced to rural areas in the early 1980s by the Local Bodies and Rural Development Department (LB&RDD) that was later renamed Local Government and Rural Development Department (LG&RDD). Assistance was provided by district and union councils with technical support from the Aga Khan Rural Support Programme (AKRSP) and financial support from the United Nations Children's Fund (UNICEF). LB&RDD formed mandatory community project committees that were responsible for providing labour and local materials, supervising construction, and managing community funds (as far as there were any). Water committees were assigned with managing the scheme post construction.

Attempts at formalizing and extending community management beyond traditional irrigation schemes can be traced back to the 1960s under President Ayub Khan (Grieser 2018). AKRSP started work in Gilgit-Baltistan in 1982, building on local institutions by formalizing existing village organizations (VOs) as a long-term institution for development, with villages forming the basic unit and their populations viewed as interest groups acting as a unified community (Clemens 2000; Khan and Hunzai 2000). A special focus on gender equality and gender equity saw the introduction of women's organizations (WOs) as women were often unable to participate in male-majority VOs due to gender segregation. VOs and WOs enabled local populations to apply to AKDN agencies for support. Since 2003, VOs and WOs have been consolidated into local support organizations (LSOs) which are hybrid professional organizations rooted in broad-based VOs and WOs (AKRSP 2011).

Table 4.1 WASEP key features, aims, objectives, and approach

Feature	*Aim*	*Overall objective*	*Specific objectives*	*Approach*
1. Social mobilization	Build capacity and involve communities in decision-making, with defined SOPs 'to impart quality and consistency'	Harness community potential through grassroots institutions for equitable and sustainable development and poverty reduction	a. Understand community dynamics b. Organize communities into COs and committees c. Develop community capacity d. Establish sustainable community-based financing mechanism for O&M	a. Interactive procedures b. Meetings and interactive methods c. Training to upgrade HR skills d. Generate, document, and audit finances (O&M fund, tariffs, new connection funds)
2. Health and hygiene (H&H) promotion	Raise awareness about personal, domestic, and environmental hygiene through participatory approaches	Improve existing H&H practices and reduce diarrhoeal morbidity	a. Help community adopt better personal, domestic, and environmental hygiene practices b. Educate the community about waterborne diseases and preventive measures c. Assist and encourage children to raise awareness of healthier behaviours and routines d. Actively involve teachers in improving H&H practices e. Ensure hygienic and proper usage of water and sanitation facilities	a. H&H education sessions in communities and schools, group discussions, and meetings b. H&H sessions, workshops, and trainings c. H&H sessions, art, and essay writing competitions d. Workshops, trainings e. Sensitize communities through household visits
3. Training and capacity development	Enable communities to successfully run WASEP projects through participatory approaches	Enhance capacities of participants to perform their respective roles in a more efficient and effective manner	a. Enable community organizations to provide demand-responsive services and participatory management b. Develop WSC leadership and managerial skills to plan, implement, and manage sustainable interventions c. Disseminate WASEP knowledge and experience among stakeholders d. Identify and build capabilities of in-house staff to keep abreast of sector developments and enhance motivation	a. Promote understanding of organization objectives through interactive and participatory process b. Share knowledge and experience through brainstorming sessions, discussions, group work, participation c. Set up aims, objectives, and intended outcomes of training d. Facilitate agreement on the delivery mechanism and implement this

(*Continued*)

Table 4.1 Continued

Feature	*Aim*	*Overall objective*	*Specific objectives*	*Approach*
4. Engineering	Successfully implement infrastructure through better understanding of relevant engineering principles and SOPs	Ensure the uninterrupted supply of safe drinking water and provision of adequate sanitation facilities to end users	a. Prepare user-friendly design of water supply and sanitation infrastructure b. Select and use quality materials and ensure proper fixing and utilization c. Provide a robust monitoring mechanism to ensure quality	a. Use global best practices and past experience b. Keep abreast of new products, use appropriately trained craftspeople c. Supervision mechanism binding senior engineers, management to regular site visits
5. Water quality management	Provide clean and safe drinking water to rural and urban communities of GBC and Sindh that meet WHO guidelines	Ensure the provision of potable water that meets WHO guidelines	a. Select source(s) for improved systems free from contaminants b. Identify water quality issues associated with existing water practices and raise community awareness of these c. Determine the degree of contamination of water sources and suggest appropriate remedial measures	a. Test all potential sources for contaminants, select alternative source, treat physical or microbiological contamination b. Analyse existing water collection points for contaminants, raise community awareness about risks c. Monitor supplied water quarterly for two years to ensure provision of agreed water quality

Note: SOP, standard operating procedure; CO, community organization; O&M, operations and maintenance; H&H, health and hygiene; WASEP, Water and Sanitation Extension Programme; WSC, Water and sanitation committee; GBC, Gilgit-Baltistan and Chitral; WHO, World Health Organization.

AKRSP also transformed the traditional meaning of communal work and community management. Traditional collective labour typically involved reciprocal work exchange (*bue*) where a household requires additional help to undertake work, and communal work (*rajaaki*) ordered by local rulers or a village representative (*nambardār*) (Hussain and Langendijk 1995; Sökefeld 1997). This background of forced labour and 'element of compulsion' meant that there was an ambivalence towards communal work (Hussain and Langendijk 1995: 9). Under AKRSP, communal work and labour was transformed into a gift to the community as opposed to the coercion of the past (Miller 2015: 33).

Social mobilization is meant to encourage participation in planning, financing, and implementation of water and sanitation schemes. WSHHSP identified a lack of wider community participation and hence ownership and cost sharing as the reasons for low levels of functionality and poor water quality. It thus emphasized the need for an even greater participatory approach involving the mobilization and participation of a majority of the project community to engender responsibility, and to include women in the entire project cycle, as women may have the greater interest as key beneficiaries (Hussain and Langendijk 1995).

The WASEP terms of partnership (ToP) requires community 'buy-in' through a number of meetings with AKAH, each with increasing attendance requirements and the endorsement of up to 100% of households at critical points of the interaction so that the project has the support of the entire village and by both women and men. Participation takes place over a series of meetings and group activities with or attended by AKAH where a minimum of 40% of women and men need to be present in the first dialogue, 50% in monthly calendar meetings, and at least 80% at the general body meeting in the final dialogue with AKAH.

WASEP's selection process seeks to balance a community's needs with its capacity to sustain a project. Although CBWM is meant to be demand led, WASEP is very much initiated and led by AKAH, and carried out by AKAH engineering, finance, and H&H teams (AKPBS-P n.d.1). Applications are invited through advertisements for the current funding phase of WASEP projects, and are shortlisted based on demonstrable needs, community capacity, and technical and economic feasibility. A pre-feasibility visit by AKAH seeks to determine the genuineness of the application and willingness of the community to accept the WASEP ToP that 'outlines roles and responsibilities and asks the villagers to commit to all portions of the programme: water supply, sanitation, drainage, operation, maintenance, and health and hygiene education' (AKPBS-P n.d.1: 2). The selection process only begins after a community has 'bought-in' and committed to the scheme by paying into an operations and maintenance (O&M) endowment fund up-front (AKPBS-P n.d.1).

AKAH's participatory approach to assess a community's ability to carry out and sustain an intervention project involves a series of participatory rural appraisals (PRAs) that employ tools such as village mapping and seasonal calendars (with separate groups for men and women) to identify specific needs of the community, the context in which WASEP would be operating, and where special considerations are needed. The 'PRA Village Selection Criteria' form allows AKAH to assess a community's needs against its capacity and covers the following six differently weighted categories:

1. Demonstrable need and availability of potential source(s) (24%): water shortage, existing water management practices, and yield of the proposed source.
2. Enabling factors (20%): presence of VO and WO, evidence of unity, financial strength (household savings), past management record/ experience.
3. Commitment to contribute (14%): acceptance of ToP, past record.
4. Cost effectiveness (24%): location of source, housing pattern, type of project, soil condition.
5. Equitable access to water and sanitation (8%): potential to cover different segments of society.
6. Existence of H&H supporting facilities (10%): schools, health facilities.

4.3 Ownership and control

WSHHSP identified problems with early project committees where participation was limited to only a minority of villagers who did not oversee technical planning, and where women were excluded (Hussain and Langendijk 1995). LB&RDD assumed that the community would take control and ownership of the project post construction, but their exclusion meant that the community did not have 'a strong sense of ownership of the scheme' or feel responsible for its management (Hussain and Langendijk 1995: 42). It was often also unclear who was responsible after project completion.

Similarly, the water management committee usually only comprised selected village notables and the union councillor with varying presence and effectiveness, was characterized by the misuse of funds with no long-term management and usually dissolved after project implementation (Ahmed and Alibhai 2000). Communities were often unable to manage O&M due to 'the absence and understanding of well defined management systems, dissatisfaction with service levels, lack of motivation, general reticence, and social conflicts' (Hussain et al. 2000: 76), continuing to rely instead on government 'to do even the small repairs which are always possible for the community to repair by themselves' (Ahmed et al. 1996: 12).

The aim of WASEP's social mobilization is thus to build capacity and involve communities in decision-making. The overall objective is to unite the community and harness its potential through the creation of grassroot institutions and a sustainable community-based financing mechanism for O&M based on an understanding of community dynamics (AKPBS-P 2012: 3). At the heart of ownership and control of water systems are the project committee that oversees implementation and the water and sanitation committee (WSC) that manages O&M. The ToP requires that the community forms a WSC, elected at a village general body meeting, to administer and manage the water supply scheme. WSCs are based on VOs and are not legal entities, with no legal documents or legal framework. They are built on trust and are part of local support organizations that are registered with AKRSP and that include VOs and WOs. WSCs comprise voluntary administrative staff (at the minimum a president, secretary, and treasurer) and two salaried employees – a water and sanitation operator (WSO or plumber) and water and sanitation implementer (WSI) for H&H awareness (AKPBS-P n.d.1; AKPBS-P 2012).

The WSC 'is responsible for administering the system' including 'planning and managing the development and upkeep of the network', setting and collecting water tariffs and all financial matters including payment of salaries, developing and enforcing local legislation vis à vis water and sanitation, and ensuring that the spare parts store is well stocked (AKPBS-P n.d.1: 5). Regular maintenance, including repairing leaks and broken faucets, is the responsibility of the WSO, who is also a member of the WSC and trained to repair valves, distribution mains, and larger components of the network. The WSO is meant to inspect the water reservoir and record the reading of the bulk flow meter, and undertake preventive maintenance (e.g. line flushing and exercising valves) (AKPBS-P n.d.1: 5). Because WASEP schemes are sometimes divided into separate projects covering different parts or neighbourhoods (*mohallah*) of a settlement, separate WASEP projects may end up sharing the same WSC. Alternatively, different WSCs around the same settlement or location may create an Apex Committee to make decisions on issues that affect all the WSCs in that location.

A key requirement for CBWM and WASEP is the community's management capacity. Training and capacity development are key to community management by enabling communities to successfully run WASEP projects through participatory approaches. The overall objective of WASEP's training and capacity development is to enhance the capacities of participants to perform their respective roles in a more efficient and effective manner by:

- enabling community organizations to provide demand responsive services and participatory management;
- developing WSC leadership and managerial skills to plan, implement, and manage sustainable interventions;
- disseminating WASEP knowledge and experience among stakeholders; and
- identifying and building capabilities of in-house staff to keep abreast of sector developments and to enhance motivation.

Training for WSCs is provided through a three-day workshop to develop the capacities for leadership and community management; record keeping and convening of meetings; maintaining accounts and bookkeeping; maintenance and repairs (for the WSO); and public speaking, demonstrating good hygiene practices and data collection (for the WSI). Training for the WSI involves a two-day in-house training and one-day exposure visit on H&H education plus a two-day refresher course. In addition, exposure visits are organized for village activists on managing issues arising, project sustainability, the tariff system and penalties, new connection policies, payment to WSOs/WSIs, and record keeping.

4.4 Cost sharing

Cost sharing in the form of regular tariff payments is central for sustaining O&M. One of the main reasons for the poor condition and non-functioning of water systems in Gilgit-Baltistan was the lack of proper O&M (Hussain et al. 2000). In common with the CBWM model, cost sharing is central to WASEP's community-based financing and is outlined in the ToP with AKAH. The community contributes to the cost of construction through in-kind contributions, generally in the form of (unskilled) labour (or the cash equivalent, usually in the case of urban projects) and local materials (e.g. sand, gravel, stone). The remainder is contributed by the external funding body, usually through a government agency and/or AKAH as the implementing NGO. AKAH also contributes expertise and non-local materials (e.g. cement, pipes, mild steel flat bars). It is worth noting that although the WASEP model of CBWM emphasizes public–private partnerships as a key feature and the way forward, in the absence of a significant private sector the 'private' component of this partnership refers either to the implementing NGO or the community rather than the (for profit) private sector. This is an important distinction insofar as AKAH's costs are lower even in comparison to government departments that might normally implement DWSS, because public–private partnerships in this case would typically involve private contractors.

The share of community contribution depends on the overall cost of the project. As a result, rural projects that are simpler and cheaper tend to have a higher share of community contributions compared to more complex and expensive (mechanized) urban projects where the community share of total project cost is lower even though the community ends up contributing more per household (see Chapter 8). The community contribution is based on a bill of quantities that calculates the quantities and cost of materials and labour. Cost sharing post-implementation is based on the community contributing to an operation and maintenance (O&M) endowment (referred to as the O&M fund), new connection fees, and monthly tariffs. The O&M fund functions as a self-managed investment fund where the profits (in the form of interest earned) contribute to the salaries of the WSO and WSI and cost of spare parts. (AKAH also contributes to the endowment fund, presumably on behalf of the funding body.)

Household contribution to the O&M fund is set at around PKR3,000 (US$19)[2] (rural) and PKR8,000 (US$50) (urban) and may vary depending on the project's requirements. The amount may be higher where operating expenses are higher, for example in the case of mechanized systems. Monthly tariffs are meant to help pay for the salaries of the WSO and WSI, and the cost of O&M including minor repairs. AKAH recommends a monthly tariff of between PKR100 (US$0.60) and PKR500 (US$3) per household depending on socioeconomic condition and location, but the final tariff is set by the community through the elected WSC.

4.5 Health and hygiene education

The main conclusion of WSHHSP was that engineering alone was not sufficient to reduce water-related diseases because there was a lack of hygiene awareness to change behaviour and cultural practices. Behavioural change is thus a major component of WASEP. WSHHSP recommended a H&H education strategy based on understanding district- and population-specific water-related problems, needs, practices, and beliefs, such as ideas and beliefs of diarrhoea transmission routes, local understandings of pathogens, and local ways of prevention and treatment (Korput et al. 1995; Langendijk et al. 1996). H&H education in particular was needed to reduce faecal- and water-related diseases (Korput et al. 1995). Community awareness of the importance of clean tap water was also seen as a motivating factor for communities to undertake small repairs themselves (Ahmed et al. 1996), protect water sources, and maintain the cleanliness and integrity of the water supply system, sanitation, and personal hygiene.

WASEP's H&H promotion thus aims to raise awareness through participatory approaches about personal, domestic, and environmental hygiene to bring about comprehensive behavioural change in everyday practices so that communities can benefit from clean water. The objectives are to improve existing H&H practices to reduce diarrhoeal morbidity by:

- helping the community adopt better personal, domestic, and environmental hygiene practices;
- educating the community about waterborne diseases and preventive measures;
- assisting and encouraging children to raise awareness of healthier behaviours and routines;
- involving teachers in improving H&H practices; and
- ensuring the hygienic and proper usage of water and sanitation facilities.

To meet these objectives AKAH conducts household visits, H&H sessions, workshops, and training in communities and schools. The three main features of WASEP's H&H education are the WSI, a Community Health Intervention Programme (CHIP), and School Health Intervention Programme (SHIP). The WSI is a trained and recognized H&H expert within the community, selected through a general body meeting and the position is designated for a woman. The WSI promotes H&H, provides support to women in terms of hygiene activity, and initially monitors H&H through bi-monthly house visits to collect information on the incidence of diarrhoea and to observe other parameters of hygiene behaviour such as latrine use and the presence of faeces around the home (AKPBS-P n.d.1: 5). The WSI also acts as AKAH's link with women and, as a member of the WSC, serves as a voice for H&H matters and for women to the committee.

CHIP and SHIP sessions aim to bring about behavioural change in terms of H&H practices to complement the provision of clean drinking water and reduce the prevalence of waterborne diseases. CHIP was initiated in conjunction with the Aga Khan University-Community Health Services to:

- create awareness about H&H;
- facilitate local action to improve domestic, personal, and environmental hygiene;
- assist villagers in sustaining adoption of healthier behaviours;
- make villagers aware that they are important partners and are responsible for improving hygienic conditions in their villages and homes;
- make partners understand how diarrhoeal diseases are spread and what preventive measures can be taken to reduce their occurrence;
- enable villagers to take required actions to cure children and other family members infected by diarrhoea; and
- ensure the use and O&M of WASEP water and sanitation schemes.

CHIP sessions are held with at least 25% of households over nine months or less, covering nine topics: (1) personal hygiene; (2) domestic hygiene; (3) environmental hygiene; (4) dirty water; (5) food and fruit handling; (6) latrine promotion and usage; (7) diarrhoea; (8) intestinal worms; and (9) transmission routes of waterborne diseases. Education sessions are led by trained female H&H promoters with the women of a village in their local languages. Although CHIP also involves men and children, the target group is women because of 'the relationship between women, water and sanitation in traditional gender roles' where they 'carry out tasks such as cooking, feeding children, and washing and cleaning' and deal with the 'disposal of wastewater, solid waste, and children's faeces' (AKPBS-P n.d.1: 6).

SHIP was introduced in 1999 in collaboration with Aga Khan University-Institute for Educational Development following observations in many programme communities that children were not washing hands after defecation or before eating, were eating unwashed fruit and drinking from open water channels, and that small children found it difficult to use and hence were not using latrines. It was recognized that sustainable behavioural change could not be achieved without addressing the educational needs of children (AKPBS-P n.d.). School children were also more susceptible to food and waterborne diseases but adopted new behaviours with less effort than adults. The objectives of SHIP are to:

- increase the H&H knowledge of children;
- assist children in making plans and taking action to create awareness and adopt healthier behaviours in their schools, homes, and villages;
- involve children in helping their siblings adopt healthier behaviours;

- make children feel that they are important and responsible actors in improving conditions and promoting hygienic practices;
- improve H&H in schools and villages; and
- actively involve teachers in the process to attain better results.

SHIP sessions are conducted with Grades 3 and 5 school children over eight months or less, covering eight topics: (1) clean hands; (2) latrine usage; (3) dirty water; (4) diarrhoea; (5) intestinal worms; (6) personal hygiene; (7) environmental hygiene; and (8) usage of tap water.

H&H education is a core component of WASEP not only because the main aim of DWSS is to reduce waterborne diseases but also as the benefits of H&H are used to encourage regular household payment for water. AKAH estimates that clean drinking water along with improved H&H behaviour through WASEP has reduced average annual household medical expenses from PKR3,070 (US$35) to PKR118 (US$1.35) (AKPBS-P n.d.). The cost of monthly regular tariff payments for WASEP water is thus argued to be much less than what households used to pay in medical expenses arising from contaminated water.

4.6 Engineering solutions

Despite the recognition that engineering solutions alone cannot deliver the health benefits of clean drinking water, engineering nevertheless remains at the core of WASEP's integrated approach and distinguishes the programme from the emphasis on 'software' or governance in the CBWM literature. Engineering design remains a central and unique feature of WASEP, with engineering principles and standard operating procedures (SOPs) seen as the basis for the uninterrupted supply of safe drinking water. This is implemented through a 'user-friendly design of water supply and sanitation infrastructure', selection and proper use of quality materials, and a robust monitoring mechanism to ensure quality (AKPBS-P 2012: 58). As mentioned earlier, in contrast to CBWM principles of community control of all decision-making, this focus on engineering also means that AKAH leads on all technical and engineering aspects of the project.

This includes: (1) selection of source(s) (based on yield, accessibility, distance, type of source, type of scheme, water quality); (2) a topographic survey (for gravity-fed schemes and mechanized systems); (3) analysis of field survey data and calculations; (4) modelling and designing the water distribution network (including calculating average and peak demands, and simulations for pipe diameters) and civil structures (intake chamber, storage tank, sedimentation tank, up-flow roughing filters, slow sand filters, valve box, distribution chamber, sump, *nallah* (stream) crossing, pump room); (5) sharing the design with the community; (6) finalizing a bill of quantities that covers labour, material, and transportation costs; (7) execution of drawings in AutoCAD; (8) site supervision; and (9) commissioning and testing (AKPBS-P 2012: 60–71).

Given the nature of these engineering tasks, and even without the constraints of community capacity, communities cannot be expected to have control or even input into every aspect of a project. Engineering design is discussed in further detail in Chapter 9.

4.7 Water quality management

Water quality management has always been key to WASEP's approach and is closely connected with its engineering focus. Engineering and environmental teams conducting research during WSHHSP on appropriate treatment methods for the region identified sedimentation and 'up-flow roughing filtration' as the most appropriate methods of reducing turbidity levels and eliminating microbial contaminants, with water treatment plants built to remove turbidity and reduce microbial contamination to World Health Organization (WHO) standards (AKPBS-P n.d.1: 4). An operational routine for the filtration units was developed 'to allow for village level maintenance and regular troubleshooting by the WSO' with AKAH providing additional training to operators for plant operation and monitoring (AKPBS-P n.d.1: 4). But AKAH also recognizes that WSCs may lack the capacity to address larger technical issues.

The aim of WASEP's water quality management is to provide clean drinking water to rural and urban communities that meets WHO standards for safe drinking water in developing countries, defined as less than 10 *E. coli* per 100 ml. This is done by: (i) carefully selecting and testing source(s) to remove contaminants; (ii) identifying water quality issues associated with existing water collection points and water practices, and raising community awareness of these; and (iii) determining the degree of contamination of water sources by monitoring the supplied water quarterly for two years to ensure the agreed water quality is provided.

Water quality monitoring occurs pre- and post-intervention with baseline source sample testing for chemical, physical, and microbiological contamination, and post-intervention tests at storage, distribution, and household. Water quality monitoring and engineering design are based on technological innovations that include software for modelling, water quality assessment equipment, and the use of high-quality parts and processes such as HDPE pipes, compression fittings, and earthquake-resistant, reinforced concrete storage tanks.

4.8 Conclusion

WASEP represents an important case study not just in its scale and delivery of clean drinking water systems (as opposed to standalone handpumps or tap stands) but also its integration of H&H awareness, social mobilization, and engineering solutions to address the problems of waterborne disease. It is also an example of the implementation of CBWM principles of participation, ownership, control, and cost sharing, and its successful delivery and continued operations of piped water systems offers important lessons for sustainable CBWM. However, WASEP is also distinct because of its integrated approach and continued emphasis of engineering or 'hardware' where the CBWM literature prioritizes management or 'software'. A further distinction is the leading role of AKAH as the implementing NGO, contrary to the CBWM requirement of community control of all aspects of water services. The following chapters examine how these features of WASEP translate into practice and if this model is sustainable and scalable.

Notes

1. This is a historical exchange rate for 1997 when the WSHHSP study was completed. Due to currency depreciation, the equivalent would be US$9.50 (in 2000) and US$3.60 (2024).
2. US dollar equivalents in the remainder of the chapter are based on average annual exchange rates for the study period of 2019–2021.

CHAPTER 5
Community participation in WASEP

Jeff Tan, Anna Grieser, Matt Birkinshaw, Saleem Uddin, and Fatima Islam

5.1 Introduction

Participation lies at the heart of community-based water management (CBWM) and the Water and Sanitation Extension Programme (WASEP). It is central to sustainable operations as it helps to identify the demand for water services and for communities to take 'ownership' of this by assuming responsibility and control. This in turn encourages the payment of tariffs to sustain operations. The ability and/or willingness of communities to pay will thus be key to sustainable CBWM. However, the demand or desire for water services does not necessarily signal the ability or even willingness of households to pay regular tariffs as illustrated by the global evidence of widespread irregular or non-payment and tariffs that are set low to be affordable but do not then cover the cost of operations and maintenance (O&M) (see e.g. Schouten and Moriarty 2003).

The evidence from WASEP largely conforms to this pattern, with most rural households not paying regular tariffs compared to urban households despite higher levels of demand, participation, and a sense of ownership (see Chapter 8). While this raises questions about the sustainability of WASEP and CBWM more generally in rural areas, the greater ability or willingness of urban households to pay suggests that WASEP may be more sustainable in urban areas.

This chapter outlines the role and main features of participation in WASEP as part of the CBWM model but that has also been facilitated by the wider social, political, and historical context (Section 5.2). It examines the evidence from a household survey of community participation in WASEP across Gilgit-Baltistan, focusing on the differences between rural and urban participation (Section 5.3), and the impact of this on the functionality and operations of water systems. It then discusses the implications of this evidence for the sustainability and scalability of WASEP (Section 5.4).

5.2 Participation in WASEP

As discussed in Chapter 1, participation in CBWM is defined as 'an active process whereby beneficiaries influence the direction and execution of development projects rather than merely receive a share of project benefits' (Paul 1987: 2, cited in Hutchings et al. 2016: 6). It encompasses a range of activities from 'information sharing (the lowest level) to consultation, decision-making, and initiating action (the highest level)' where 'the community performs routine operational duties such as record keeping, accounting, and payment collecting under a system predefined by an external agency' (McCommon et al. 1990: 7, 10). Defined as such, participation in CBWM centres very much on community input and management and operations, through the active and regular participation in the decision-making process. This is conceptually distinct from what often appears as community participation (or community action) (see e.g. Schouten and Moriarty 2003) through in-kind (unskilled) labour contributions that are mainly payments made in the absence of cash contributions.

Community participation is based on the idea of the community as culturally and politically homogeneous and harmonious. While this has been shown to not be the case even in villages (see Hutchings et al. 2015), rural communities can nonetheless be expected to be more homogeneous and cohesive, thereby facilitating social mobilization and participation compared to more diverse, recent, and transient populations that characterize towns and cities. The level of participation will also depend on socioeconomic conditions and the historical context, including specific non-project aspects of people's lives and their complex livelihood inter-linkages (Cleaver 1999). Very low-income communities for example only have low to moderate levels of participation compared to middle- and high-income communities (McCommon et al. 1990). Conversely, communities that have a history or prior experience of participation, for example working with NGOs that employ participatory approaches, can be expected to have higher levels of participation than those that have not. Urban water systems also tend to be public compared to many rural communities that still own irrigation systems and who are used to the self-reliant practice of building and managing water schemes and resources that is missing in urban settings. The expectation that water systems should be public and free further undermines urban participation, particularly when politicians use free water as political leverage to secure votes.

In the case of WASEP, successful rural participation can be traced back to historical and socio-cultural factors, in particular: (a) the introduction of participatory approaches in Gilgit-Baltistan in the 1960s under President Ayub Khan's development programme; (b) the formation of the Aga Khan Rural Support Programme (AKRSP) initiative and its introduction of formalized community structures referred to as village organizations (VOs) in the 1980s; and (c) community-based projects under the Local Bodies and Rural Development Department (LB&RDD) from the early 1980s (Birkinshaw et al. 2021). WASEP thus benefited from earlier work on social mobilization by AKRSP and other Aga Khan Development Network (AKDN) agencies as well as earlier government-implemented, community-based projects as part of the wider promotion of participatory approaches.

Capacity building and strengthening of local institutions remains an important aspect of WASEP's participatory approach, and the implementing NGO, the Aga Khan Agency for Habitat (AKAH), usually works with previously established VOs and women's organizations. The role of women's participation is discussed separately in Chapter 6. In contrast, most old and many new urban settlers (especially in Gilgit City) come from districts that have no, or only a recent, history of AKDN participatory engagement and are thus less likely to participate. Active involvement in project committees also requires time and effort that urban residents are less likely to have.

As described in Chapter 4, WASEP is very much led by AKAH, which publicizes each new programme or phase based on the availability of external funding and invites applications from communities in targeted or priority areas. Applications are meant to signal the desire or demand for improved water, sanitation, and hygiene (WASH) services, and the willingness and capacity of communities to participate and contribute to the implementation and management of WASEP as per the programme's terms of partnership (see Chapter 2). Based on interviews with local social mobilizers, 16.7% of the sample rural VOs obtained information about WASEP from visits by AKAH and other AKDN agencies. However, the largest proportion (40%) of rural villages inquired directly with AKAH about WASEP, with 33.3% the result of word of mouth and another 16.7% from the 'demonstration effect' of seeing (successful) WASEP schemes in other villages.

The WASEP model considers participation of the whole community as central and its standard operating procedures specify how communities need to be involved at different stages of the project (Chapter 4). A project committee is typically formed for the provision of labour and local materials, supervision of construction, and collection and management of community funds. The wider community is expected to participate in meetings for: planning and implementation (with AKAH), (village) discussions on WASEP, a mapping exercise (as part of a needs assessment by AKAH), health and hygiene training (by AKAH), water and sanitation committee (WSC) selection, and regular post-implementation discussions with the WSC.

Community-based financing is an important part of WASEP's participatory approach and provides the link between participation and sustainability. This centres on an O&M (endowment) fund based on the initial payment of each participating household along with regular (monthly) water tariffs and new connection fees (AKPBS-P 2012) (see Chapter 8). Once the O&M fund is collected, implementation requires labour contributions. These are usually carried out by men with women's participation in digging taking place in only a few communities and otherwise restricted to encouraging, and catering food for, male labourers (see Chapter 6). However, changing livelihoods along with urbanization have meant that urban households are less inclined to contribute in-kind labour and instead opt to pay for this work to be done by hired labour, which potentially undermines the idea of community participation.

WASEP's participatory approach aims to establish a sense of ownership which in turn is expected to encourage regular financial contributions for long-term operations, management, and financing of WASEP schemes. The actual management usually only involves a few (male) WSC members rather than the active engagement of the community. Most WSC members are not paid for their time although some may hope for indirect rewards such as status, social capital or even the chance to be employed by an AKDN organization in future (Grieser 2018). Such dedicated individuals usually play a decisive role in convincing the majority of the community of the worth of a WASEP project, submitting, and following up on, the project application, mobilizing full community participation, and facilitating process-oriented decision-making (Grieser 2018).

5.3 Household survey responses

Evidence of participation in WASEP is drawn from questions around social (community) mobilization (SM) in the household survey. These questions seek to establish the level of desire (SM1) and hence demand for water services, the level of participation at various stages of implementation (SM2–SM5) and post implementation (SM27, SM28), and the sense of responsibility (SM15) as a proxy for (sense of) ownership (Table 5.1).

The data for Gilgit-Baltistan indicates overall high levels (79%) of desire for improved WASH (SM1), with higher demand most notably among rural (90.2%) compared to urban (72.7%) households (Table 5.2). This is partly mirrored in higher demand in homogeneous projects (82.8%) (defined as where 90% or more of households come from the same sectarian group) compared to heterogeneous projects (77.1%) (where less than 90% come from the same sectarian group). Communities categorized as 'unknown' are mixed communities where information on actual shares of sectarian groups is uncertain but are most likely heterogeneous. The average overall participation (SM2–SM5, SM27–SM28) is higher in homogeneous communities (49.5%) compared to heterogeneous (36.3%) and unknown (45.4%) communities (Table 5.2).

However, the key difference is between rural and urban participation where rural participation is noticeably higher during planning and implementation (SM2), group discussions (SM3), mapping with AKAH (SM4), and WSC selection (SM5). The average total participation (SM2–5, SM27–28) is higher for rural (58.5%) compared with urban (35.5%). Overall higher levels of rural participation are also reflected in a higher sense of responsibility for WASEP (51.6%) (SM15) compared to urban households (35.2%). Nonetheless, overall participation levels and sense of responsibility for WASEP are much lower than the desire for WASH (SM1), indicating that actual participation does not match demand for WASH.

Table 5.1 Household survey questions: participation

Code	*Topic*	*Question*	*Answer options*
SM1	Desire for improved WASH	Before WASEP, how much desire was there in your household for improvements in water, sanitation, and health in your village/area?	Very little Not much Some Quite a lot A lot Don't know
SM2	Participation (WASEP planning, implementation)	How much were you or someone from your household involved in the discussions during WASEP project planning and implementation?	Not at all Only a little Somewhat Quite a lot Very much Don't know
SM3	Participation (group discussion on WASEP)	How many group discussions on the WASEP project did a member of your household attend?	1, 2, 3 ... 7 8 or more None Don't know
SM4	Participation (mapping with AKAH)	Did you or a member of your household take part in your village/area mapping, calendar, history, and illness discussions with AKAH?	Yes No Don't know
SM5	Participation (WSC selection)	Did you or a member of your household participate in deciding on the members of the water project's WSC in your village/area?	Yes No Have not heard of this decision-making process Don't know
SM27	Participation (WSC meeting last six months)	In the last six months have you or someone from your household participated in a group meeting on the current or future work of the water project's WSC?	Yes No Don't know
SM28	Participation (village/area meeting last six months)	In the last six months, have you or someone from your household participated in a meeting on the current or future work of any of these groups or organizations in your village/area (multiple answers possible)	Yes (total comprising: village meeting of all residents; meeting with AKAH staff; village organization; women's organization; WSC; *Jamat Khana*; mosque; village council)
SM15	Responsibility for WASEP	How much responsibility do take for the WASEP project?	Very little Not much Some Quite a lot A lot Don't know

Note: WASH, water, sanitation and hygiene; WASEP, Water and Sanitation Extension Programme; AKAH, Aga Khan Agency for Habitat; WSC, water and sanitation committee; Jamat Khana (congregational place)

Higher levels of rural participation are very likely the function of demographic differences, specifically community diversity that is also reflected in the number of languages spoken, with rural WASEP projects being culturally less diverse (more homogeneous) and tending to comprise longer-term residents. Participation levels are lower in larger, more mixed communities with a greater proportion of recently arrived residents. In urban projects, the number of languages spoken is strongly and significantly negatively correlated with the level of participation, although in rural projects there is no significant relationship. If projects are divided into groups by the number of households, the negative correlation between the number of languages and participation is only statistically significant for large projects.

Participation in WSC meetings in the last six months (SM27) across all districts is much lower than participation during planning and implementation (SM2), group discussions (SM3), mapping (SM4), and WSC selection (SM5), with Gilgit (13.2%) and Diamer (15.6%) having the lowest participation (Table 5.3). The difference in participation in WSC meetings is also much smaller between rural (23.2%) and urban (16.1%) (Table 5.2). Higher levels of rural (63.8%) and urban (61.4%) participation in village/area meetings over the last six months (SM28) indicate continued general participation in comparison to rural (23.2%) and urban (16.1%) participation in WSC meetings (Table 5.2). These household responses suggest declining participation in water management that could be because water services have been successfully delivered and management has subsequently been assigned to the WSC. Wider participation in this case appears to be restricted to the planning and implementation phase of WASEP, with the WSC then expected to manage and operate the water scheme.

These results are broadly replicated across the eight districts, with Gilgit district (where the majority of WASEP urban projects are located) having the lowest desire for improved WASH (71.2%) compared to the average across all eight districts (88.6%) (Table 5.3). Similarly, the average total participation for Gilgit district (34.5%) is much lower than the district average (54.0%), with household participation in WASEP planning and implementation (SM2), group discussions on WASEP (SM3), mapping with AKAH (SM4), WSC selection (SM5), and WSC meetings in the last six months (SM27) all substantially lower (Table 5.3). High average overall participation levels are reflected in higher levels of responsibility for WASEP (SM15) in Astore (51.7%), Diamer (62.5%), Ghizer (50.5%), Hunza (50.9%), and Nagar (51.7%) compared with Gilgit district (33.9%).

Data from Jutial and Danyore schemes in Gilgit City account for the lower levels of participation in Gilgit district. The average desire for WASH across the eight urban Jutial projects (67.3%) is lower than for Gilgit district (71.2%) with the highest desire in Noorabad Extension (75.9%) and Astore Colony (75.0%) and lowest in Wahdat (53.7%) (Table 5.4). Average participation levels for Jutial projects as a whole (32.8%) are slightly lower than Gilgit district (34.5%). There are nonetheless variations among Jutial projects, with the

Table 5.2 Participation survey responses: Gilgit-Baltistan rural, urban (%)

	Gilgit-Baltistan	*Rural*	*Urban*	*Homogeneous*	*Heterogeneous*	*Unknown*
Sample size (no.)	*3,132*	*1,125*	*2,007*	*1,426*	*1,194*	*512*
SM1 Desire for improved WASH						
Very little	5.6	2.4	7.4	4.4	5.9	8.4
Not much, some	9.4	5.8	11.4	8.4	10.2	10.4
Quite a lot, a lot	79.0	90.2	72.7	82.8	77.1	72.9
SM2 Participation (WASEP planning, implementation)						
Not at all	23.0	10.5	30.0	20.1	28.1	19.3
Only a little, somewhat	26.0	22.7	27.8	23.1	28.7	27.6
Quite a lot, very much	42.8	63.6	31.1	51.3	32.0	44.2
SM3 Participation (group discussion on WASEP)						
Eight or more meetings	15.8	27.5	9.2	21.5	8.0	17.8
1–7 meetings	47.8	56.7	42.9	49.1	46.4	47.7
None	26.4	11.0	35.0	22.5	33.2	29.7
SM4 Participation (mapping with AKAH)						
Yes	36.0	56.0	24.8	44.8	26.1	34.4
No	51.3	37.6	59.0	46.1	58.8	48.4
Don't know	10.6	5.0	13.7	7.4	12.6	14.6
SM5 Participation (WSC selection)						
Yes	39.2	60.4	27.4	46.6	33.1	41.2
No	49.0	34.4	57.1	45.4	51.9	44.3
Don't know	7.9	2.9	10.7	5.0	10.3	9.4

(Continued)

Table 5.2 Continued

	Gilgit-Baltistan	*Rural*	*Urban*	*Homogeneous*	*Heterogeneous*	*Unknown*
Sample size (no.)	*3,132*	*1,125*	*2,007*	*1,426*	*1,194*	*512*
SM27 Participation (WSC meeting last six months)						
Yes	18.6	23.2	16.1	17.9	19.3	25.6
No	57.9	55.7	59.1	60.9	55.5	53.5
Don't know	8.1	4.8	9.9	5.0	10.7	11.7
N/A	15.4	16.2	14.9	16.3	14.6	9.2
SM28 Participation (village meeting last six months)						
Yes	62.2	63.8	61.4	65.7	59.8	61.7
(Meeting with AKAH staff)	6.2	1.9	3.9	2.5	1.9	6.1
(WSC)	7.8	5.3	2.0	4.6	2.5	10.5
No	14.0	23.1	9.0	16.8	11.7	12.1
Don't know	20.8	10.0	26.9	15.1	25.7	23.2
SM15 Responsibility for WASEP						
Very little	12.4	9.7	13.8	11.6	13.0	14.1
Not much, some	37.1	30.5	40.7	35.3	38.5	34.4
Quite a lot, a lot	41.1	51.6	35.2	42.6	39.8	40.8
Don't know	6.4	5.6	6.9	7.0	5.9	7.4
Average overall participation	**43.7**	**58.5**	**35.5**	**49.5**	**36.3**	**45.4**

Note: WASH, water, sanitation, and hygiene; WASEP, Water and Sanitation Extension Programme; AKAH, Aga Khan Agency for Habitat; WSC, water and sanitation committee

Table 5.3 Participation survey responses: Gilgit-Baltistan districts (%)

	Astore	*Diamer*	*Ghizer*	*Gilgit*	*Hunza*	*Nagar*	*Shigar*	*Skardu*
Sample size	*60*	*32*	*614*	*1,703*	*369*	*58*	*162*	*134*
SM1 Desire for improved WASH								
Very little	1.7	0.0	3.1	8.3	3.0	0.0	1.2	0.7
Not much, some	10.0	0.0	5.9	11.0	11.7	1.7	8.0	6.7
Quite a lot, a lot	88.3	100.0	89.1	71.2	83.5	98.3	89.5	88.8
SM2 Participation (WASEP planning, implementation)								
Not at all	13.3	9.4	7.5	32.7	14.9	10.3	17.9	12.7
Only a little, somewhat	18.3	28.1	23.8	27.2	23.0	20.7	27.8	31.3
Quite a lot, very much	68.3	62.5	66.4	27.9	56.6	62.1	51.9	49.3
SM3 Participation (group discussion on WASEP)								
Eight or more meetings	26.7	18.8	33.6	7.2	22.8	20.7	12.3	19.4
1–7 meetings	58.3	71.9	54.2	41.0	49.6	55.2	66.0	64.9
None	6.7	0.0	8.6	38.5	16.5	22.4	17.9	7.5
SM4 Participation (mapping with AKAH)								
Yes	63.3	43.8	57.5	22.4	43.1	53.4	44.4	59.0
No	30.0	56.3	36.8	61.7	39.8	39.7	51.2	30.6
Don't know	6.7	0.0	4.7	13.0	15.7	6.9	3.1	7.5
SM5 Participation (WSC selection)								
Yes	66.7	62.5	64.0	24.4	48.2	55.2	53.1	48.5
No	26.7	37.5	31.6	60.0	39.0	41.4	42.0	40.3
Don't know	5.0	0.0	2.6	10.7	8.1	3.4	3.1	6.7

(*Continued*)

Table 5.3 Continued

	Astore	*Diamer*	*Ghizer*	*Gilgit*	*Hunza*	*Nagar*	*Shigar*	*Skardu*
Sample size	*60*	*32*	*614*	*1,703*	*369*	*58*	*162*	*134*
SM27 Participation (WSC meeting last six months)								
Yes	33.3	15.6	24.6	13.2	27.4	20.7	22.8	24.6
No	55.0	84.4	52.9	61.8	49.3	51.7	59.9	50.0
Don't know	10.0	0.0	3.6	9.2	13.8	10.3	2.5	5.2
N/A	1.7	0.0	18.9	15.7	9.5	17.2	14.8	20.1
SM28 Participation (village meeting last six months)								
Yes	70.0	62.5	62.9	70.9	61.2	63.8	79.0	62.7
(Meeting with AKAH staff)	13.3	0.0	6.2	6.5	5.4	0.0	8.0	6.0
(WSC)	16.7	25.0	18.9	1.4	10.3	25.9	0.0	0.0
No	8.3	37.5	28.8	8.3	16.5	36.2	7.4	7.5
Don't know	18.3	0.0	5.9	28.7	19.5	0.0	9.3	22.4
SM15 Responsibility for WASEP								
Very little	3.3	6.3	11.2	14.0	9.2	22.4	7.4	12.7
Not much; some	35.0	21.9	33.1	41.2	29.5	13.8	38.9	35.8
Quite a lot, a lot	51.7	62.5	50.5	33.9	50.9	51.7	42.0	46.3
Don't know	10.0	9.4	3.3	6.9	7.9	12.1	9.9	1.5
Average overall participation	**64.4**	**56.3**	**60.5**	**34.5**	**51.5**	**55.2**	**54.9**	**54.7**

Note: WASH, water, sanitation, and hygiene; WASEP, Water and Sanitation Extension Programme; AKAH, Aga Khan Agency for Habitat; WSC, water and sanitation committee

Table 5.4 Participation survey responses: Jutial (%)

	Aminabad	*Astore Colony*	*Diamer Colony*	*Noor Colony*	*Noorabad Extension*	*Wahdat Colony*	*Yasin Colony*	*Zulfiqarabad*
Sample size	*77*	*68*	*70*	*80*	*79*	*82*	*72*	*82*
SM1 Desire for improved WASH								
Very little	2.6	2.9	8.6	13.8	12.7	12.2	13.9	2.4
Not much, some	14.3	11.8	17.1	7.5	8.9	19.5	18.1	7.3
Quite a lot, a lot	72.7	75.0	61.4	63.8	75.9	53.7	61.1	74.4
SM2 Participation (WASEP planning, implementation)								
Not at all	35.1	27.9	35.7	23.8	41.8	58.5	26.4	42.7
Only a little, somewhat	32.5	25.0	12.9	23.8	29.1	19.5	34.7	23.2
Quite a lot, very much	23.4	30.9	37.1	45.0	25.3	12.2	25.0	29.3
SM3 Participation (group discussion on WASEP)								
Eight or more meetings	20.8	11.8	4.3	5.0	3.8	11.0	6.9	19.5
1–7 meetings	33.8	32.4	21.4	47.5	50.6	24.4	38.9	28.0
None	40.3	39.7	60.0	36.3	43.0	58.5	33.3	41.5
SM4 Participation (mapping with AKAH)								
Yes	23.4	35.3	20.0	27.5	11.4	14.6	20.8	23.2
No	66.2	50.0	58.6	57.5	81.0	78.0	48.6	64.6
Don't know	9.1	11.8	18.6	15.0	6.3	4.9	25.0	11.0
SM5 Participation (WSC selection)								
Yes	26.0	35.3	15.7	26.3	15.2	17.1	11.1	23.2
No	59.7	52.9	70.0	61.3	77.2	72.0	55.6	68.3
Don't know	7.8	5.9	11.4	11.3	5.1	4.9	23.6	7.3

(*Continued*)

Table 5.4 Continued

	Aminabad	*Astore Colony*	*Diamer Colony*	*Noor Colony*	*Noorabad Extension*	*Wahdat Colony*	*Yasin Colony*	*Zulfiqarabad*
Sample size	*77*	*68*	*70*	*80*	*79*	*82*	*72*	*82*
SM27 Participation (WSC meeting last six months)								
Yes	13.0	27.9	2.9	31.3	6.3	8.5	2.8	1.2
No	84.4	52.9	72.9	57.5	88.6	73.2	69.4	79.3
Don't know	2.6	5.9	12.9	11.3	3.8	3.7	6.9	4.9
N/A	0.0	13.2	11.4	0.0	1.3	14.6	20.8	14.6
SM28 Participation (village meeting last six months)								
Yes	88.3	82.4	21.4	57.5	65.8	84.1	75.0	70.7
(Meeting with AKAH staff)	18.2	17.6	0.0	8.8	8.9	12.2	2.8	0.0
(WSC)	3.9	2.9	1.4	0.0	0.0	7.3	1.4	1.2
No	0.0	4.4	7.1	0.0	0.0	0.0	2.8	6.1
Don't know	11.7	11.8	68.6	42.5	32.9	12.2	16.7	23.2
Responsibility for WASEP								
Very little	15.6	4.4	5.7	7.5	12.7	15.9	22.2	20.7
Not much, some	39.0	42.6	32.9	12.5	57.0	39.0	51.4	37.8
Quite a lot, a lot	39.0	32.4	30.0	58.8	26.6	32.9	18.1	23.2
Don't know	3.9	2.9	28.6	20.0	0.0	6.1	2.8	18.3
Average overall participation	**38.1**	**42.6**	**20.5**	**40.0**	**29.7**	**28.7**	**30.1**	**32.5**

Note: WASH, water, sanitation, and hygiene; WASEP, Water and Sanitation Extension Programme; AKAH, Aga Khan Agency for Habitat; WSC, water and sanitation committee

highest participation in Astore Colony (42.6%) and Noor Colony (40.0%) and lowest in Wahdat (28.7%), Noorabad Extension (29.7%), and Yasin Colony (30.1%). Households with 'quite a lot' and 'a lot' of sense of responsibility for WASEP (SM15) are also lowest in Noorabad Extension (26.6%) and Yasin Colony (18.1%), along with Zulfiqarabad (23.2%), and account for the overall lower share in Jutial (32.6%) compared to the district (33.9%) and urban (35.2%) averages. Low levels of participation in WSC meetings over the last six months (SM27) in Gilgit district (13.2%) can be attributed to projects in Jutial (11.7%) and in particular Diamer Colony (2.9%), Yasin Colony (2.8%), and Zulfiqarabad (1.2%). This is despite high levels of participation in (general) village/area meetings in the last six months (SM28) for Jutial projects overall (68.2%) that again suggests lower community interest and participation post implementation or after the WSC takes over.

Such findings may be explained by intra- and inter-community factors. In the case of Noorabad Extension, extraordinary engagement by some community members to collect funds and demonstrate the community's willingness and capacity to work on the project also shortened the social mobilization process which may help explain low participation levels. In Noor Colony and Aminabad, inter-community rivalry for funds (for the first explicitly 'urban' project) may have intensified social mobilization as communities had to not only demonstrate that there was a need but also the financial and management capacity. This may explain high levels of participation in Noor Colony and Aminabad. In Yasin Colony, residents had registered a desire for a WASEP project 10 years ago, but the neighbourhood was too small and would have had to merge with Diamer Colony whose residents were not considered by Yasin Colony residents as ready to engage with such a project (Grieser 2018).

WASEP peri-urban projects in Danyore (located on the outskirts of Gilgit City) offer a useful comparison to urban projects in Jutial. Given its settlement history, participation levels of households in Danyore can be expected to be closer to that of rural households compared to Jutial. The average desire for WASH across the nine Danyore projects (76.4%) is higher than for Gilgit district (71.2%) and noticeably higher than Jutial (67.3%). The highest levels of desire are in Syedabad (93.7%), Shangote Patti (82.6%), and Hunza Patti (81.5%), and the lowest level in Sharote etc. (55.7%) (Table 5.5). Average participation levels for Danyore (32.3%) are similar to Jutial (32.7%) but conceal a much wider range as reflected in lower median participation levels of 25.3% compared to 32.7% for Jutial, with the lowest level of participation in Amphary Patti (16.8%). This only partly corresponds with fewer households that felt 'quite a lot' and 'a lot' of responsibility for WASEP in Amphary Patti (16%) and in Danyore projects as a whole (33.5%). Higher participation in WSC meetings in the last six months (SM27) for Danyore projects (13.6%) compared to Jutial (11.7%) and Gilgit district (13.2%) may be because the Danyore scheme was still under construction, although this figure is still lower than the urban average (16.1%).

Table 5.5 Participation survey responses: Danyore (%)

	Amphary Patti	*Chikas Kote*	*Hunza Patti*	*Hussainpura etc.*	*Princeabad Bala etc.*	*Shangote Patti*	*Sharote etc.*	*Sultanabad 1&2*	*Syedabad*
Sample size	*106*	*89*	*92*	*102*	*91*	*121*	*88*	*119*	*79*
SM1 Desire for improved WASH									
Very little	5.7	4.5	1.1	1.0	14.3	2.5	11.4	12.6	0.0
Not much, some	2.8	15.7	15.2	10.8	7.7	7.4	25.0	12.6	3.8
Quite a lot, a lot	75.5	69.7	81.5	77.5	76.9	82.6	55.7	74.8	93.7
SM2 Participation (WASEP planning, implementation)									
Not at all	46.2	20.2	29.3	10.8	30.8	39.7	21.6	42.0	36.7
Only a little, somewhat	13.2	42.7	27.2	34.3	33.0	24.8	33.0	37.0	25.3
Quite a lot, very much	10.4	23.6	38.0	31.4	29.7	21.5	34.1	18.5	27.8
SM3 Participation (group discussion on WASEP)									
Eight or more meetings	1.9	0.0	0.0	5.9	3.3	3.3	1.1	5.0	8.9
1–7 meetings	25.5	59.6	71.7	45.1	52.7	36.4	61.4	38.7	36.7
None	53.8	23.6	23.9	20.6	36.3	45.5	25.0	53.8	38.0
SM4 Participation (mapping with AKAH)									
Yes	12.3	10.1	45.7	21.6	18.7	18.2	31.8	15.1	22.8
No	69.8	62.9	47.8	52.9	67.0	61.2	55.7	76.5	58.2
Don't know	10.4	25.8	3.3	20.6	14.3	17.4	12.5	8.4	15.2
SM5 Participation (WSC selection)									
Yes	8.5	24.7	37.0	22.5	29.7	19.8	37.5	17.6	22.8
No	75.5	55.1	58.7	51.0	58.2	58.7	46.6	67.2	58.2
Don't know	3.8	13.5	4.3	21.6	11.0	14.9	9.1	11.8	15.2

(Continued)

Table 5.5 Continued

	Amphary Patti	*Chikas Kote*	*Hunza Patti*	*Hussainpura etc.*	*Princeabad Bala etc.*	*Shangote Patti*	*Sharote etc.*	*Sultanabad 1&2*	*Syedabad*
Sample size	*106*	*89*	*92*	*102*	*91*	*121*	*88*	*119*	*79*
SM27 Participation (WSC meeting last six months)									
Yes	1.9	7.9	23.9	19.6	4.4	23.1	14.8	10.1	16.5
No	36.8	24.7	58.7	60.8	45.1	55.4	30.7	83.2	75.9
Don't know	12.3	24.7	7.6	14.7	8.8	18.2	4.5	5.0	3.8
N/A	49.1	42.7	9.8	4.9	41.8	3.3	50.0	1.7	3.8
SM28 Participation (village meeting last six months)									
Yes	40.6	38.2	58.7	65.7	37.4	65.3	51.1	73.1	81.0
(Meeting with AKAH staff)	2.3	5.9	13.0	0.0	8.8%	16.5	11.1	8.4	7.8
(WSC)	0.0	0.0	1.9	0.0	2.9	1.3	0.0	0.8	3.1
No	28.3	28.1	2.2	0.0	24.2	0.0	28.4	0.0	0.0
Don't know	21.7	33.7	39.1	30.4	38.5	33.9	18.2	25.2	17.7
SM15 Responsibility for WASEP									
Very little	22.6	4.5	17.4	2.9	14.3	13.2	21.6	22.7	3.8
Not much, some	39.6	65.2	28.3	47.1	33.0	41.3	67.0	42.0	53.2
Quite a lot, a lot	16.0	28.1	41.3	41.2	49.5	42.1	9.1	32.8	41.8
Don't know	12.3	2.2	13.0	2.0	3.3	2.5	1.1	0.8	0.0
Average overall participation	**16.8**	**27.3**	**45.8**	**35.3**	**29.3**	**31.3**	**38.6**	**29.7**	**36.1**

Note: WASH, water, sanitation, and hygiene; WASEP, Water and Sanitation Extension Programme; AKAH, Aga Khan Agency for Habitat; WSC, water and sanitation committee

While overall sect is an important predictor of participation and responsibility levels, variation in participation and responsibility levels in Danyore can be attributed to internal political conflict around the role of the Government of Gilgit-Baltistan (GoGB) that had agreed to partially fund a WASEP project for the greater Danyore area. However, local politicians sought to secure votes by arguing that GoGB should fund the entire project thereby exempting household cash, labour, and O&M contributions. This resulted in adherents of different political parties and neighbourhood and language groups engaging differently during implementation. In some neighbourhoods, households delayed payment in the hope for a court decision that the water supply would be the sole responsibility of GoGB and not the community (see Chapter 7). Participation levels in these neighbourhoods are likely to have been lower than in other neighbourhoods where community representatives and residents increased their efforts and financial contributions to complete the project as initially agreed between the community representatives, AKAH, and GoGB. It is also very likely that lower levels of initial participation, communication, and a full understanding of WASEP might have resulted in lower approval of the project in some neighbourhoods, resulting in lower overall participation levels and willingness to contribute.

5.4 Community participation, sustainability, and scalability

The significance of community participation depends on whether this positively impacts sustainability and whether the scaling up of WASEP to urban areas such as Jutial and Danyore can be replicated. Sustainability can be measured in terms of functionality (or continued operations) and financial performance. As the evidence shows, participation in WASEP projects is determined by location (rural/urban) more so than social composition (homogeneity/heterogeneity) and is higher in rural projects. The continued operation of all but 1 of the 25 sampled rural projects demonstrates the ability of rural communities to sustain operations, with a third of the functional schemes operating at or beyond their 15-year life cycle, and another third operating for around 10 years or more (see Chapter 9).

This is despite a third of rural WSCs not collecting regular tariffs and 12 (48%) not employing a water and sanitation operator (WSO or plumber) for regular maintenance (see Chapter 8). Instead, communities in the majority of rural projects have opted to contribute to maintenance and repairs as and when the need arises. These community cash (and in-kind) contributions have been important in sustaining operations including undertaking repairs to damage from natural hazards. Of the 12 rural projects in the engineering sub-sample (Chapters 9 and 10), all were disrupted by natural hazards, and all except one (Chandupa) were able to mobilize labour and/or cash contributions to undertake repairs, often with help from external agencies (see Chapter 10).

This appears to be regardless of participation levels, with communities in four public (non-WASEP) control sites in Ghizer, Gilgit, Hunza, and Skardu also able to mobilize communities to pay for and/or undertake repairs (Chapter 10). In the case of WASEP projects, communities with both high participation levels in Daeen Chota (62.1%), Dushkin (64.0%), Kirmin (77.8%), and Nazirabad (60.9%) and lower levels of participation in Hatoon Paeen (45.6%), Rahimabad (Matumdass) (37.4%), and Singul Shyodass (49.4%) were all able to undertake repairs. Similarly, participation levels do not appear to be correlated with WSC performance as captured by household survey questions on ratings for the WSO, WSC, water quantity/quality, and user satisfaction with the cost of WASEP water.

Nonetheless, estimates of operating revenue and costs suggest that rural projects may not be financially sustainable (beyond the 15-year lifespan of water infrastructure and given the need for system rehabilitation and expansion). Due to low or absent tariffs and irregular tariff payments, only 6 (24%) of 24 functioning rural WSCs were estimated to have monthly operating surpluses (see Chapter 8). Of the eight rural projects that did not have monthly or regular tariff collections, households in seven had opted to pay only when repairs were needed. Households in one project (Broshal) stopped tariff payments, believing that their money had been embezzled, and opted to collect tariffs on a needs basis. (AKAH later clarified that this was simply a case of the WSO withdrawing PKR30,000 (US$190 based on the average exchange rate for the research period in 2019–2021) that he had deposited for the whole community to secure the position of WSO but then quit and withdrew his money when he realized the salary was too low.) Households in Chandupa stopped paying because the WSC was inactive and the project stopped functioning in 2010. When asked about the introduction of water meters, eight (32%) WSCs reported that households were too poor, and seven (28%) stated that there was no support for this. High levels of rural participation are thus not reflected in attitudes towards, or the payment of, regular tariffs, consistent with the wider evidence (Chapter 1).

In contrast, urban households with lower levels of participation demonstrate a greater willingness to pay, even if urban tariffs are considerably higher. (In Noor Colony, for example, the community financed an extensive system replacement that had become necessary due to initial miscalculations.) As a result, urban projects performed better financially despite costing more and requiring higher tariffs. This may be because urban households do not have alternative sources of water and need to support the project. It is also likely that higher incomes allow urban households to pay higher and more regular tariffs. This greater ability and willingness to pay appears to be separate from participation levels and indicates that sustainability may depend on the financing and management of water services rather than community participation.

Participation is also central to the management of WSCs and can be assessed in terms of the willingness of communities to stand for WSC elections (as per the WASEP terms of partnership) and how often WSC positions are rotated. Regular WSC elections and the replacement of committee members can indicate a willingness of communities to participate in the management of water systems, although one senior AKAH manager pointed out that WSC turnover may be due to dissatisfaction and that non-participation is a reflection of project success as the service was being delivered. WSC membership is usually determined by elections, with the community voting as a whole, or by community members volunteering. It was noted that there was a wider problem of participation and specifically with replacing WSC members, with 76% rural and 62.5% urban WSC positions relying on community volunteers rather than elections. This is reinforced by the same committee members being re-selected in rural (84%) and urban (81.3%) WSCs, which mirrors the wider evidence of the limits of 'informality and voluntarism' in CBWM (see e.g. Hutchings et al. 2015). This is consistent with recommendations for the professionalization of community management (see e.g. World Bank 2017) with at least one urban WASEP WSC employing a manager.

5.5 Conclusion

Participation is at the heart of CBWM and WASEP because this is meant to encourage ownership and hence the regular tariff payments necessary for sustainable operations. The introduction and focus of CBWM in the countryside leverages the idea of rural communities being more likely to practise self-help and to participate in the delivery and management of water services. This is supported by the evidence from the household survey that clearly shows higher rural compared to urban participation levels that correspond with the degree of community diversity. Rural communities are more homogeneous than urban communities and this is reflected in higher participation levels in rural and, to a lesser extent, homogeneous WASEP projects. Participation rates in WASEP are also higher compared to control sites, suggesting that social mobilization and the earlier work of AKDN agencies play a part in encouraging participation. Higher levels of participation correlate with more successful WASEP projects as measured by: household scores for water quality and quantity (Chapter 9); conflict (unity, collective/harmonious water decisions, arguments over water, and fair allocation) (Chapter 7); attitudes to tariffs (Chapter 8); and WSC performance.

However, the sustainability of WASEP and CBWM is predicated on participation increasing the sense of ownership and in turn encouraging the regular payment of tariffs for O&M. The evidence shows that participation levels are much lower than the demand (or desire) for WASH, and more critically, do not correspond with the ability or willingness of households to pay. This is because rural households tend to have lower incomes and are less able or willing to pay regular tariffs, especially if income sources are irregular or seasonal (see e.g. Whaley and Cleaver 2017). Rural households in the WASEP sample are engaged more in agriculture or work in the military compared to urban households that are represented more in white collar employment and business. Interestingly urban respondents reported only slightly higher levels of educational attainment when asked about the highest level of education in the household.

Not surprisingly then, urban households with lower levels of participation demonstrate a greater willingness to pay regular tariffs due to higher levels of income and the lack of alternative sources of water. Danyore projects were not yet operational at the time of the research. But the large variation in participation rates both in Jutial and Danyore projects illustrates that some communities are more likely to participate than others. However, these varying levels of participation do not appear to correspond with operational and financial performance and hence sustainability. In the case of Jutial, projects with the highest (Astore Colony, Noor Colony) and lowest levels of participation (Wahdat, Noorabad, Yasin Colony) all have regular monthly collections of tariffs and are all estimated to also have monthly operating surpluses (Chapter 8). This suggests that management and the willingness to pay are more critical than participation in ensuring sustainability as illustrated by the fall in household participation in WSC meetings post implementation.

CHAPTER 6
Women's participation and WASEP

Anna Grieser, Jeff Tan, Matt Birkinshaw, Saleem Uddin, Yasmin Ansa, and Karamat Ali

6.1 Introduction

A central aspect of community-based water management (CBWM) is the participation of all community members, including women. As primary carers, women are considered the main beneficiaries and thus the main proponents of (clean) water supply schemes (Cleaver 1998; Oakley 1991; Uphoff et al. 1998; Van Wijk-Sijbesma 1997, 1998; Whaley and Cleaver 2017). Women's involvement in planning, operations, management, and decision-making processes is expected to lead to better performance, sustainability, and lower costs (Cleaver 1998; Carrard et al. 2013; Fisher 2006; van Wijk-Sijbesma 2001), not least because they have a vested interest in the success and sustainability of water services. However, there is little evidence of the link between participation, including women's participation, and improved performance and sustainability, and furthermore women continue to be under-represented, especially at leadership positions, with low participation levels in CBWM (see e.g. Hannah et al. 2021).

The Water and Sanitation Extension Programme (WASEP) is based on the premise that women have a greater desire and hence incentive than men for the introduction of drinking water supply schemes (DWSS) because the burden of fetching, storing, and using water in rural households falls primarily on women (Hussain and Langendijk 1995). The participation of women is seen to have positive outputs both on short-term project implementation and the long-term goal to improve the lives, health, and hygiene of local populations. This is because of women's roles as carers who are responsible for health and hygiene (H&H) awareness raising, which is a key component of WASEP's integrated approach. Women can thus be seen as active participants in pushing for the implementation of WASEP in their village and in bringing about H&H improvements through behavioural change so that communities can reap the full benefits of clean drinking water.

However, the degree of women's participation will be context specific, and in particular, dependent on the wider sociocultural context. This is especially the case for rural areas that tend to be socially more conservative, and more so in places such as Gilgit-Baltistan and Pakistan more generally, where religion, tradition, and customs tightly specify the role and place of women. Not surprisingly perhaps, and consistent with the wider evidence, women's participation in WASEP is low and much lower than general (village-wide) participation levels during the planning and implementation phases. It is even lower in meeting attendance post-implementation and the actual management of WASEP. Even (paid) positions earmarked for women as water and sanitation implementers (WSIs) to promote H&H education have been largely unfilled. This raises questions about the likelihood of increasing women's participation and representation in the wider context of prevailing social norms and resource constraints.

This chapter examines the argument for women's participation generally and in WASEP (Section 6.2) before assessing the level of this participation in WASEP, looking at the evidence from the household survey, interviews with water and sanitation committees (WSCs), and focus group discussions (FGDs) (Section 6.3). It then discusses whether women's participation has made, or can make, a difference in WASEP, and the implications of this for sustainability and scaling up (Section 6.4).

6.2 Women's participation and sustainability

The desirability of women's participation in water management is based on women being more knowledgeable and better managers of water and funds than men since they are usually the main water users (Cleaver 2000; UN-DESA 2005). However, the evidence is mixed, with women's participation shown to strongly enhance project effectiveness and sustainability (Mommen et al. 2017; Nishimoto 2003, cited in Were et al. 2008) but also not necessary for successful community participation (Narayan 1995, cited in Prokopy 2004; Prokopy 2004; Ray 2007). At the same time, women continue to be under-represented or excluded in irrigation (see e.g. Khandker et al. 2020; Zwarteveen 2008) and domestic water, where 'there is a mismatch between the women's domestic water needs and the governance of domestic water services' (Mandara et al. 2017: 188). Even where women are part of management committees (as a result of national policies requiring a quota for women), they often do not participate or are represented by male kin (Cleaver 1991; Meinzen-Dick and Zwarteveen 1998; Prokopy 2004; Singh 2008; Were et al. 2008; Zwarteveen and Neupane 1996), with participation of women in community managed projects in general usually much lower than that of men and much lower than expected (Schnegg and Linke 2016: 812; Mandara et al. 2017; Were et al. 2008).

The Aga Khan Agency for Habitat (AKAH), the implementing NGO for WASEP, believes that women's participation in all phases of a project can only be beneficial, not just for short-term project implementation and management but also for the long-term goal of improving the lives, health, and hygiene of local populations. This belief that women are crucial for long-term sustainability has been reinforced by subsequent studies reviewing early WASEP projects (Alibhai et al. 2001; Barnett et al. 2001). Women's participation in water and sanitation projects in Gilgit-Baltistan is, however, constrained by a number of factors such as local gender roles (including a gendered division of labour), women's lack of access to education and information, and leadership positions dominated by men (Halvorson 1994, cited in Halvorson et al. 1998).

WASEP thus incorporates a number of institutional mechanisms to encourage women's participation and improve planning, implementation, and maintenance of DWSS. This includes separate project planning meetings with women and female AKAH staff to collect gender-specific data and incorporate women's concerns in communities where women are free to join meetings with other women. AKAH staff are instructed to maximize women's participation in all phases of a project including application, planning, hardware selection, implementation, and system management.

Female AKAH staff also conduct Community Health Intervention Programme (CHIP) sessions with female community members, in addition to School Health Intervention Programme (SHIP) sessions at local schools, employing participatory methods to convey health education and hygiene training to women and children (Halvorson et al. 1998). Communities are encouraged to engage a woman as a salaried WSI to examine the community's H&H conditions, and instruct on personal, domestic, and environmental hygiene (Alibhai et al. 2001; Ahmed and Alibhai 2000; AKPBS-P and USAID 2016; AKF and AKPBS-P 2014). The WSI is also supposed to be part of the WSC post-handover and to represent women and their concerns on the WSC. AKAH staff members also encourage communities to engage women in tariff collection (Alibhai et al. 2001; Hussain et al. 2000). However, it is not clear if the WSI position is meant to be permanent and there is no official documentation of this. As the aim of the WSI is to assist in the promotion of H&H awareness, this would not have been expected to continue indefinitely once the project was completed, and AKAH expected the WSI to work for an average of four to five years.

6.3 Women's participation in WASEP

Evidence of women's participation comes from the household survey questions on women's participation (SM6) and participation in women's organization meetings as part of village meetings in the last six months (SM28) (Table 6.1). These are part of wider questions on social mobilization (SM) and participation (Chapter 5) focused on the role of women. This is supplemented by interviews with WSC members (including current and former women WSIs) and FGDs with community and WSC members.

For SM6 (women's participation in activities that led to WASEP), 47.1% of households in Gilgit-Baltistan reported that women did not participate at all compared to 20.8% for 'quite a lot, very much' (Table 6.2) and 22.5% for 'only a little, somewhat'. Given the history of women's organizations in villages,

Table 6.1 Household survey questions: women's participation

Code	*Topic*	*Question*	*Answer options*
SM6	Participation (women)	How involved were women in your village/area in activities that led to WASEP?	Not at all Only a little Somewhat Quite a lot Very much Don't know
SM28	Participation (village meeting last six months)	In the last six months, have you or someone from your household participated in a meeting on the current or future work of any of these groups or organizations in your village/area?	Yes (women's organization) No Don't know

Table 6.2 Women's participation survey responses: Gilgit-Baltistan (GB), districts (%)

	Participation: not at all	*Participation: quite a lot, very much*	*Meeting last six months: women's organization*	*Social composition*	*Majority sect*
Province (GB):					
Overall	47.1	20.8	0.6	–	–
Rural	32.5	36.2	0.1	–	–
Urban	55.2	12.3	0.5	–	–
HOM	42.1	28.8	0.1	–	–
HET	51.3	14.2	0.0	–	–
District:					
Astore	68.3	16.7	0.0	HOM	Sunni
Diamer	96.9	0.0	0.0	HOM	Sunni
Ghizer	28.0	37.9	0.2	HET	Ismaili
Gilgit	57.7	9.6	0.8	HET	Mixed
Hunza	21.7	44.4	1.1	HOM	Ismaili
Nagar	50.0	36.2	0.0	HOM	Shia
Shigar	54.9	19.1	0.0	HOM	Shia
Skardu	38.1	23.1	0.0	HOM	Shia

Note: HOM, homogeneous; HET, heterogeneous

meaningful women's participation ('quite a lot, very much') is, as would be expected, significantly higher among rural (36.2%) compared to urban (12.3%) households, with a slightly smaller but still significant gap between homogeneous (28.8%) and heterogeneous (14.2%) settlements. In both cases, over half of women in urban (55.2%) and heterogeneous (51.3%) households did not participate at all, and women's participation in general is around half that of overall rural and urban household participation in Gilgit-Baltistan (see Chapter 5, Table 5.2).

There are, however, large variations in women's participation across the eight districts of Gilgit-Baltistan related to different local practices, ranging from the highest in Ghizer (37.9%), Hunza (44.4%), and Nagar (36.2%) to the lowest in Astore (16.7%), Diamer (0%), and Gilgit district (9.6%) (Table 6.2). The large percentage of non-participation of women in Astore (68.3%) and Diamer (96.9%) corresponds to largely homogeneous, socially conservative populations in both districts, while non-participation in Gilgit (57.7%) is consistent with the high number of urban projects and hence lower urban participation levels generally. Lower women's participation is also related to concerns over safety in towns with diverse non-kin populations.

Higher rural participation levels for women thus conceal wide variations between districts and are also reflected in the sample of 25 rural WASEP projects. This ranges from the highest levels of women's participation in Diruch (87.2%), Hundur Barkulti (89.5%), and Kuno (89.7%), all homogeneous communities with Ismaili majorities, to the lowest in Burdai (0%), Shilati (0%), and Sutopa (8.3%), homogeneous communities with Shia or Sunni majorities (Table 6.3). This variation is also reflected in the higher average level of women's participation in Ismaili settlements (44.9%) compared to Shia (18.5%) and Sunni (7.0%) (the caveat being that there were only two homogeneous Sunni majority settlements in the rural sample).

Low average participation levels for women in Gilgit district can be traced to very low women's participation in Jutial (9.4%) and Danyore (9.5%) (Table 6.4). The low women's participation in Jutial is due to Astore Colony (2.9%), Diamer Colony (4.3%), Noor Colony (6.3%), Noorabad (2.5%), and Wahdat (1.2%), almost all of which are homogeneous and all of which consist of 80–100% Sunni populations. The average level of women's participation in Jutial is higher in Ismaili-majority settlements (20.7%) compared to Sunni (3.4%). In the case of Danyore, women's participation is low overall regardless of the majority sect, with the exception of Hussainpura etc. (18.6%) and Sharote etc. (20.5%).

Meetings over the last six months for women's organizations was negligible in Gilgit-Baltistan (0.6%), with higher participation in urban (0.5%) and homogeneous (0.1%) compared to rural (0.1%) and heterogeneous (0%) households (Table 6.2). This suggests that households in some homogeneous communities may have higher women's participation levels. This is supported by the district level data where the only participation in women's organizations is in Ghizer (0.2%) and Hunza (1.1%), both Ismaili-majority districts, along with Gilgit (0.8%) (Table 6.2). Women's participation in Gilgit district is due to meetings in a handful of sites in Danyore: Hussainpura etc. (4.5%), Shangote Patti (1.3%), Sharote etc. (2.2%) – all of which are Shia majority settlements – and Sultanabad 1&2 (1.7%) which is Ismaili majority (Table 6.4).

Table 6.3 Women's participation survey responses: rural sample (%)

	Participation: not at all	*Participation: quite a lot, very much*	*Meeting last six months: women's organization*	*Social composition*	*Majority sect*
Broshal	50.0	36.2	0.0	HOM	Shia
Burdai	20.0	0.0	0.0	HOM	Shia
Chandupa	37.9	22.7	0.0	HOM	Shia
Daeen Chota	36.4	15.9	0.0	HET	Ismaili
Diruch	2.1	87.2	0.0	HOM	Ismaili
Dushkin	70.2	14.0	0.0	HOM	Sunni
Duskhore Hashupi	43.3	20.0	0.0	HOM	Shia
Halpapa Astana	63.6	18.2	0.0	HET	Shia
Hasis Paeen	20.0	37.1	0.0	HOM	Ismaili
Hatoon Paeen	22.6	28.3	0.0	HOM	Ismaili
Hundur Barkulti	2.6	89.5	0.0	HOM	Ismaili
Hyderabad Center	23.5	37.3	2.0	HOM	Ismaili
Janabad	9.5	57.1	0.0	HOM	Ismaili
Kirmin	2.2	71.1	0.0	HOM	Ismaili
Kuno	2.6	89.7	0.0	HOM	Ismaili
Marikhi	40.7	22.2	0.0	HOM	Ismaili
Nasir Abad	22.9	18.8	0.0	HOM	Ismaili
Nazirabad	44.8	34.5	0.0	HOM	Ismaili
Rahimabad Mdass	44.8	10.3	0.0	HET	Ismaili
Shamaran Paeen	24.4	46.7	0.0	HOM	Ismaili
Shilati	96.9	0.0	0.0	HOM	Sunni
Singul Shyodass	31.0	27.6	3.4	HOM	Ismaili
Staq Paeen	64.0	24.0	0.0	HOM	Shia
Sutopa	80.6	8.3	0.0	HOM	Shia
Yuljuk	31.3	31.3	0.0	HOM	Norbakshi

Note: HOM, homogeneous; HET, heterogeneous

Evidence from the household survey is supported by interviews with WSC members and (former) WSIs. Women's participation in the WSC is guaranteed with the inclusion of the WSI which is reserved for women. However, while 21 (80%) of the 25 rural WSCs sampled had women WSIs at the start, only four WSCs (16%) (Hasis Paeen, Kirmin, Shamaran Paeen, Yuljuk) had a (woman) WSI at the time of the field visit, with only one of these positions (Shamaran Paeen) being paid a salary (Table 6.5). Shilati in Diamer district had a male WSI due to restrictions on women's mobility. Of the remaining rural WSCs, five (20%) (Hatoon Paeen, Kuno, Marikhi, Nasir Abad, Nazirabad) secured the services of a Lady Health Worker (LHW), a community

Table 6.4 Women's participation survey responses: urban sample (Jutial, Danyore) (%)

	Participation: not at all	*Participation: quite a lot, very much*	*Meeting last six months: women's organization*	*Social composition*	*Majority sect*
Jutial:					
Aminabad	37.7	16.9	0.0	HET	Mixed
Astore Colony	69.1	2.9	0.0	HOM	Sunni
Diamer Colony	82.9	4.3	0.0	HOM	Sunni
Noor Colony	81.3	6.3	0.0	HET	Sunni
Noorabad Ext	65.8	2.5	0.0	HOM	Sunni
Wahdat Colony	90.2	1.2	0.0	HOM	Sunni
Yasin Colony	37.5	9.7	0.0	HOM	Ismaili
Zulfiqarabad	37.8	31.7	0.0	HOM	Ismaili
Average	62.8	9.4	0.0		
Danyore:					
Amphary Patti	57.5	8.5	0.0	HET	Mixed
Chikas Kote	51.7	4.5	0.0	HET	Ismaili
Hunza Patti	82.6	9.8	0.0	HET	Ismaili
Hussainpura etc.	30.4	18.6	4.5	HET	Shia
Princeabad Bala etc.	67.0	4.4	0.0	HET	Shia
Shangote Patti	64.5	4.1	1.3	HET	Shia
Sharote etc.	28.4	20.5	2.2	HET	Shia
Sultanabad 1&2	58.0	5.9	1.7	HET	Ismaili
Syedabad	57.0	8.9	0.0	HET	Ismaili
Average	55.2	9.5	1.1		

Note: HOM, homogeneous; HET, heterogeneous

health worker position paid for by the state to perform a similar role. Evidence from FGDs indicates that most rural WSCs do not consider the duties of WSIs important, with male members of one WSC questioning the need to pay a salary for this position. This is reflected in low WSI salaries, WSIs leaving the position early, and an unwillingness of women to take up the position.

In the 16 completed urban projects, only three WSCs currently have a WSI (Aliabad Centre, Aminabad, Rahimabad Aliabad), all of whom receive a salary. Aliabad Centre and Rahimabad Aliabad share the same WSC and the WSIs also function as tariff collectors and are paid a commission for meeting tariff collection targets on top of their WSI salaries (Table 6.6). Just under half (12) of urban WSCs had a WSI at the start, of which six WSCs no longer have this position. Five WSCs never had a WSI (Astore Colony, Diamer Colony, Soni Kot, Wahdat Colony, and Yasin Colony) and three projects with the same WSC (Chokoporo Gahkuch Bala, Domial Buridur Gahkuch Bala, Khanabad Gahkuch Bala) secured the services of an LHW (see Table 6.6).

Table 6.5 Women's participation: WSI position and WSC meetings, rural

	WSI start	*WSI current*	*WSI paid*	*WSI years*	*WSC meetings**	*WSC men*	*WSC women*
Broshal	1	No	No	4	Never	12	1
Burdai	1	No	No	1.5	No data	8	0
Chandupa	1	No	No	8	Never	6	0
Daeen Chota	1	No	No	No data	No data	4	0
Diruch	1	No	No	0.2	No data	6	1
Dushkin	0	No	No	0	No data	7	0
Duskhore Hashupi	0	No	No	0	No data	14	2
Halpapa Astana	1	No	No	2	Never	14	0
Hasis Paeen	1	Yes	No	13	Never	9	2
Hatoon Paeen	1	LHW	No	1	No data	7	0
Hundur Barkulti	0	No	No	0	Never	18	0
Hyderabad Center	1	No	No	No data	No data	18	4
Janabad	1	No	No	2	On invitation	8	2
Kirmin	1	Yes	No	No data	Sometimes	7	1
Kuno	1	LHW	No	4	Never	6	4
Marikhi	1	LHW	No	4	On invitation	10	0
Nasir Abad	1	LHW	No	3	Previously	6	2
Nazirabad	1	LHW	No	3	Sometimes	7	1
Rahimabad Mdass	1	No	No	No data	No data	6	1
Shamaran Paeen	1	Yes	Yes	13	Regularly	18	4
Shilati	1	No	No	No data	No data	5	0
Singul Shyodass	1	No	No	4	Never	6	0
Staq Paeen	1	No	No	No data	No data	8	0
Sutopa	0	No	No	0	No data	7	1
Yuljuk	1	Yes	No	10	Previously	N/A	N/A

Note: *WSI attendance in WSC meetings; LHW, Lady Health worker

Women's participation in rural WSCs was very low, with 11 (45.8%) of the 24 WSCs with data not having any women members, and women's representation on the remaining 13 WSCs ranging from 7.7% (Broshal) to 40.0% (Kuno) (average 17.4%, median 14.3%) (Table 6.5). Of the 14 rural WSCs with WSI interviews, only one (Shamaran Paeen) involved the regular participation of a WSI at WSC meetings. Women's participation in urban WSCs was lower, with 25% in Aliabad Centre, 14.3% (Aminabad) and 12.5% (Rahimabad Aliabad), and the WSI only attending WSC meetings 'when needed' (Table 6.6).

Although women's participation and empowerment are part of WASEP and often a donor requirement, these are not enforced because of cultural

Table 6.6 Women's participation: WSI position and WSC meetings, urban

	WSI start	*WSI current*	*WSI paid*	*WSC meetings**	*WSI years*	*WSC men*	*WSC women*
Aliabad Centre	2	Yes	Yes		7	6	2
Aminabad	1	Yes	Yes	When needed	8	6	1
Amphary Patti	N/A	No	No		N/A	6	
Astore Colony	0	No	No		0	6	
Chikas Kote etc.	N/A	No	No		N/A	6	0
Chokoporo Gahkuch Bala	1	LHW	No		9	6	0
Diamer Colony	0	No	No		0	3	0
Domial Buridur Gahkuch Bala	1	LHW	No			6	0
Hassan Abad Aliabad	1	No	No	Never	2	6	0
Hunza Patti	N/A	N/A	No		N/A	6	0
Hussainpura etc.	N/A	N/A	No		N/A	7	0
Khanabad	1	LHW	No			7	0
Noor Colony	1	No	No			8	0
Noorabad Extension	2	No	No			5	0
Princeabad Bala etc.	N/A	N/A	No		N/A	7	0
Rahimabad Aliabad	1	Yes	Yes		7	8	1
Sakarkoi	1	No	No			7	0
Shangote Patti	N/A	N/A	No		N/A	7	0
Sharote etc.	N/A	N/A	No		N/A	7	0
Soni Kot	0	No	No			6	0
Sultanabad 1&2	4	No	No			12	0
Syedabad	N/A	N/A	No		N/A	6	
Wahdat Colony	0	No	No			5	0
Yasin Colony	0	No	No			7	0
Zulfiqarabad	2	No	No			8	0

Note: *WSI attendance in WSC meetings; LHW, Lady Health Worker

sensitivities and AKAH's primary mandate to deliver drinking water supply systems, with water works generally viewed as men's work. As a result, women's participation in decision-making and management processes through WSCs is limited and restricted mainly to the paid position of WSI. But this position is often not filled (for reasons discussed earlier) and not permanent in most projects, with women ceasing to participate once the position is no longer funded, usually when the WSC or community decides not to pay for this through monthly tariff collections.

However, participation can take many forms (Prokopy 2004) and water works often follow local gender-specific patterns (Were et al. 2008). As such, women can participate in decision-making in formal institutions such as the WSC or contribute in informal ways which may sometimes go unnoticed (Meinzen-Dick and Zwarteveen 1998), even if patterns of participation replicate wider gender norms. Data from FGDs suggests that women's participation or contribution are less in the implementation process such as attending meetings or participatory rural appraisals and more in influencing decisions and supporting men's work from the privacy of the home. This includes convincing male household members of the worth of a WASEP project and motivating them to pay to complete the labour work (mentioned in 13 of 31 FGDs) and catering for labouring men (16 FGDs). These complementary tasks are seen locally as part of the implementation of the scheme with women directly involved through gendered work (Bourque and Warren 1981, cited in Zwarteveen 2008; Were et al. 2008). However, while labour contributions such as digging (i.e. 'men's work' or *mardōn ka kām*) are explicitly referred to in official documents and even assigned a monetary value in project costs, women's contributions tend to go unnoticed or unmentioned in official documents such as AKAH records.

This is also reflected in local perceptions of women's participation in the household survey where a significant minority of respondents downplayed women's participation because this may not always be visible even within the community. FGD data show that women contribute to the household's share of money for the operations and maintenance (O&M) fund. Women's engagement may also go unnoticed by men or in the wider community if this does not take place in public. In the case of Hunza Patti (Danyore), 82.6% of household respondents said that women were 'not at all' involved in the activities that led to WASEP and only 9.8% mentioned 'quite a lot' or 'a lot'. A gender breakdown of respondents showed that more men respondents (88.3%) believed that women participated 'not at all' compared to women respondents (71.9%). More crucially, 21.9% of women respondents believed that women participated 'quite a lot' compared to 3.3% of men respondents. Evidence from the FGDs suggests that women were the 'driving force' in securing the WASEP project, asking a political candidate contesting for a seat in the Gilgit-Baltistan Legislative Assembly for a safe drinking water project during the 2015 election campaign. Upon winning the seat and becoming a cabinet minister, the politician secured Government of Gilgit-Baltistan funding for the WASEP urban scheme for Danyore.

6.4 Women's participation, sustainability, and scalability

The question of whether women's participation improves the performance of community management and hence the sustainability of WASEP depends on the extent and nature of women's participation. Women's involvement in planning, operations, management, and decision-making processes is expected to lead to better performance, sustainability, and lower costs.

However, the evidence shows very low levels of women's participation. This can be understood in terms of the prevailing social context including religious or cultural practices and gendered relations around work and water. This means that women's participation is often informal and not in planning, operations, management, or decision-making.

Women are also said to have a vested interest in the success and sustainability of water services as key stakeholders and beneficiaries, but this is again not consistently reflected in the evidence. A breakdown of responses by gender shows a slightly greater desire for improved water, sanitation, and hygiene (WASH) by women respondents (83.7% 'quite a lot' and 'a lot') compared with men respondents (78.4%) but disaggregating the data by location shows that fewer rural women respondents (82.2%) had 'quite a lot' and 'a lot' of desire for improved WASH compared to rural men respondents (90.9%). Women's greater vested interest in the success and sustainability of WASEP is evident in urban areas where 83.9% of women respondents had 'quite a lot' and 'a lot' of desire for improved WASH compared to 68.4% of urban men respondents. Both rural (15.7%) and urban (13.8%) women believed that women participated 'quite a lot' or 'very much' compared to rural (10.8%) and urban (10.7%) men. The premise that women have a greater desire and hence incentive than men for the introduction of DWSS (Hussain and Langendijk 1995) and that this desire will translate into women's participation is thus only partially borne out by the evidence. Responses to household survey questions on women's time use also show that time saved by the introduction of piped water was used for increased domestic or agricultural work. If improved water supply does not noticeably lead to improved quality of life for women, they may not necessarily support it.

One measure of women's participation is its impact on user satisfaction as a measure of project performance over time. To assess the relationship, a 'satisfaction score' indicator measuring user perceptions of project outcomes was created. It consists of the percentage of respondent households reporting 'high' or 'very high' levels of satisfaction that the WASEP project 'was worth their hard work and payments', 'high' or 'very high' levels of fairness in WASEP water allocation, and 'high' or 'very high' ratings for WSC performance (see Birkinshaw et al. 2021). The household survey data does not show significant relationships between women's involvement in WASEP and user satisfaction. There is also no significant statistical correlation between either women's initial participation or long-term WSC participation and success, as measured by water scores, conflict scores, attitudes to tariffs, and WSC scores.

In the urban data there is a positive relationship (high female participation, higher mean satisfaction) but this is not statistically significant. There is no relationship in the rural data. This can be interpreted to mean that both women's participation and satisfaction levels are overall lower in urban projects. The apparent relationship in the aggregate data simply reflects the differences between urban and rural, not a relationship between women's participation and satisfaction levels. The relationship between women's involvement and satisfaction can also be expressed with linear models. For household level data, women's participation has a small positive effect (0.36) on satisfaction levels, while being an urban project has a much larger and more significant negative effect (–8.71). The model appears to explain 19% of the variation in satisfaction levels.

For project level data, when the urban or rural variable is included in the model, high and low levels of women's involvement show no statistically significant effects on mean satisfaction levels. Satisfaction levels also do not appear to be influenced by the presence or absence of a WSI. Models show very minor effect of the WSI years of service (0.09, significant at $p < 0.1$). Women's participation is significantly correlated with frequent payments to the WSC. However, high women's participation also correlates to lower payments, and frequency of payments does not correlate with user satisfaction. This may reflect higher women's participation in some rural areas with higher general participation and more successful projects.

When disaggregated by sect, women's participation shows mixed relationships with household satisfaction. For rural Shia, Shia-majority, and Sunni project households, and for urban Ismaili- and Shia-majority project households, a greater level of women's participation in the project was significantly correlated with lower levels of satisfaction with the project. For urban Ismaili-majority, mixed and unknown sect project households, greater levels of women's participation in the project is significantly and positively related with satisfaction. Overall, it is difficult to substantiate a link between satisfaction, sustainability, and women's participation since many factors determine whether a project fails or can be sustained (Hemson 2002: 25) and whether and how much the users will appreciate its services.

6.5 Conclusion

Women's involvement in community water management is generally considered necessary to improve performance and sustainability, and to lower costs. At the same time, women are expected to want to participate in management and decision-making because they have the most to gain from successful and sustainable water services. However, women's participation is generally low because of the local context. The case of WASEP helps illustrate some of these constraints. Although the participation of women is central and strongly encouraged in WASEP, it is also invariably constrained by the wider social context, specifically prevailing social norms (including the seclusion of women) and local beliefs and traditions that water infrastructure is the domain of men. One reason why women engage less overall in water projects is because men in Gilgit-Baltistan have traditionally been responsible for water infrastructure (planning, construction, operation, maintenance, and repair) and irrigation (Grieser 2018; Hewitt 1989; Kreutzmann 1988; Schouten and Moriarty 2003).

In the case of Gilgit-Baltistan, social norms are also determined by religious and cultural beliefs and practices that in turn vary among the different sectarian groups. Women's participation is thus lowest where the majority sects tend to be conservative in districts such as Astore and Diamer, rural settlements such as Dushkin (Astore), Shilati (Diamer), and Sutopa (Skardu), and even in both homogeneous and heterogeneous urban settlements such as Astore Colony, Diamer Colony, Noor Colony, Noorabad Extension, and Wahdat Colony (all in Jutial). Conversely, higher women's participation can be found where the majority sect is less socially conservative in districts such as Ghizer and Hunza, rural settlements such as Diruch (Ghizer), Hundur Barkulti (Ghizer), Kirmin (Hunza), and Kuno (Ghizer), and urban neighbourhoods in Gilgit City such as Zulfiqarabad and Sharote etc.

However, overall women's participation in urban projects is generally very low regardless of sectarian composition because public spaces in urban centres are predominantly populated by strangers and not kin unlike in villages, thus restricting women's mobility (Besio 2007; Gratz 2006). In the sites where the WSI position was never implemented, WSC members had argued that there was no need for (repeated) H&H education and monitoring, or that their community was socially and culturally too diverse to allow women out in public. In the case of the latter, a strong purdah culture in urban sites noted by AKAH is also present in some rural sites; the position of the WSI was never implemented in Dushkin (Astore), Duskhore Hashupi and Sutopa (Shigar), and Hundur Barkulti (Ghizer) – all homogeneous Sunni or Shia settlements.

At the same time, the inability or unwillingness of communities to pay regular tariffs that in part constitute the salary for the WSI has discouraged women from taking up this position. The discontinuation of the WSI position in rural sites due to the non-payment of salaries was mentioned 12 times in interviews with (former) WSIs, followed by marriage or educational migration (mentioned three times), low community interest (twice), and low WSI commitment (once). In urban sites, discontinuation was due to marriage and educational migration (mentioned seven times), non-payment of salaries (three times), low community interest (three times), and the WSI feeling uncomfortable about moving through the neighbourhood (twice). Given the household work and responsibilities of women, there is little incentive and support in the household for women to participate in unpaid work in contrast to men's voluntary engagement.

The emphasis and ongoing promotion and even requirement of women's participation without taking into account these wider social constraints thus often precludes meaningful women's participation, and risks being tokenistic (see e.g. Prokopy 2004). This is especially if women are not trained and remuneration in the case of WSI salaries are dependent on tariff payments. The inability to fill the WSI position is directly related to women having to leave for paid work elsewhere, while low WSC membership is symptomatic of wider gender relations that cannot easily be altered by (externally imposed) gender requirements. The starkest illustration of this is the male WSI position in Shilati (Diamer) because social norms meant that women were not allowed out in public.

As the WSI is often the sole woman representative on the WSC, the discontinuation of this position usually means the absence of women's participation in most WSCs. Although WASEP is based on the assumption that women's participation at the different stages of the project will benefit the community, AKAH staff believe that schemes will not be adversely affected without women's participation in the case of H&H practices as these can be implemented once a scheme is operational. This is because the overall design of WASEP is seen to be informed by extensive studies on women's needs and preferences. Finally, there does not appear to be a correlation between women's participation and sustainability or scalability.

CHAPTER 7

Water-related conflict and conflict management in WASEP

Jeff Tan, Anna Grieser, Stephen Lyon, Matt Birkinshaw, and Saleem Uddin

7.1 Introduction

Community-based water management (CBWM) often centres on the idea of community as essentially harmonious and homogeneous, 'bounded by geographical links, such as a village, settlement or district, politics or natural boundaries but also ... brought together by lifestyle, culture, religion, hobby and interest' (Wasonga et al. 2010: 167, cited in Broek 2017). There is now recognition that such simplistic and romantic notions of (rural) communities 'ignore the messy reality that no rural community is the same due to culture, religion, history and population and that differences also exist within communities caused by wealth, gender, ethnicity or religion' (Broek 2017: 29).

The challenge of CBWM thus centres on the ability not just to mobilize the community but to also manage and resolve any conflict that invariably arises during the planning, implementation, and post-implementation phases of a project. Water-related conflict affects costs and hence sustainability by delaying implementation and completion, and increasing fixed and recurring costs in cases where mechanized systems are needed because access to closer water sources is denied. Sustainability then depends on the ability of communities, and in particular water and sanitation committees (WSCs), to mediate and resolve conflicts. Given that many communities lack (conflict) management capacities, external support will usually be needed to build these capacities.

This chapter looks at water-related conflict and conflict management in Water and Sanitation Extension Programme (WASEP) schemes as a key condition for sustainable CBWM. In Gilgit-Baltistan this conflict typically centres on traditional water rights and access to water sources, particularly between old and new settlers in urban areas. As outlined in Chapter 4, the WASEP model requires evidence of unity and the absence of conflict over water rights and source selection for a village to qualify. This initially led to the selection of villages where Aga Khan Development Network (AKDN) agencies have had a presence, with early WASEP schemes often benefiting from the prior work of AKDN with specific communities in the region.

This chapter examines the differences within communities and whether more homogeneous (typically rural) communities are more cohesive and harmonious than more diverse (typically urban) communities, and the impact of this on overall unity (or the absence of conflict) that is important for the successful and sustainable management of water services. Section 7.2 presents an overview of water rights and sources of water-related conflict in Gilgit-Baltistan. This provides the wider context to examine water-related conflict and conflict resolution in WASEP in Section 7.3. Evidence of conflict and conflict resolution will be drawn from Aga Khan Agency for Habitat (AKAH)/WASEP reports, a household survey of over 3,000 rural and urban households, interviews, and focus group discussions (FGDs).

7.2 Water rights and sources of water conflict in Gilgit-Baltistan

Conflict over water rights in Gilgit-Baltistan needs to be understood in terms of the organization of communities around hydro-social systems or hydraulic units. These encompass natural resources, water supply infrastructure (small canals and channels), local communities (that build, maintain, and use these), and specific rules, rights, and obligations for distribution, operations, and maintenance. Water rights and rules define the main canals, distribution ratios, specific rotation systems between the rights holders, and responsibilities for repair and maintenance (Hill 2012). The allocation of water is usually divided between different clans and families that have typically (but not always) been those that participated in the construction of canals (Grieser 2018; Kreutzmann 2000a).

The rights to water and other natural resources (e.g. land, stream or *nallah* water, high pastures, and forests) have been fixed over multiple generations in relation to specific irrigated land and families who are responsible for rendering the land irrigable. Hydraulic units are thus characterized by specific individuals or families who can derive claims to rights over the natural resources as descendants of families that built the existing water infrastructure (Grieser 2018; Kreutzmann 2000a, 2000b). The inclusion of individual new settlers in many localities has generally been handled quite flexibly over time, where individual migrants usually either married into a rights-holding family or settled on barren land (*nautōṛ/ḵhālisa*) and were included in the existing irrigation system (Sökefeld 1997). Under Kashmiri and British rule in the 19th and 20th centuries, the rights to natural resources of the area, including *nallah* water, high pastures, forests, and adjacent barren land, were fixed (*wājib ul-arz*) in many regions, based on previous customs. Water rights (*huqūq-e ābpāshi*) or water customs (*riwāj-e ābpāshi*) were similarly fixed, and thus specific families possess water rights, usually dependent on the possession of irrigable, taxed land (*ābādī zamīn*) and the participation in repair and maintenance works of the channels (Grieser 2018; Sökefeld 1997). These may be either oral or written and signed arrangements and it is also common that written knowledge is passed on orally (Hill 2012: 26). While the physical infrastructure may be similar, water rights and irrigation rules are locally specific (Kreutzmann 2000b).

From the 1950s, with large numbers of new settlers moving to Gilgit City, residents of the city started a social movement, developing criteria to determine 'original' settlers who could genuinely claim customary rights not only to water, but to all the natural resources of and near a hydraulic unit. The main criterion was having possessed irrigable, taxed land before 1947. Increasingly 'old settlers' (*pushtūne bāshinde*) were distinguished from other (new) settlers (*bāshinde*). Since then the right to use channel infrastructure and *nallah* water, which is especially crucial in spring when water is still scarce, along with adjacent high pastures and forests, and the right to claim adjacent barren land (*shāmilāt/dās/k̲hālisa*), lies with those who can claim to be old settlers. Old settlers can however make exceptions to this and grant *nallah* water from their own share to new settlers (Grieser 2018; Sökefeld 1997, 1998). Similar distinctions between old and new settlers seem to have been applied across Gilgit-Baltistan since then.

Common conflicts are over irrigation water and specifically intra-unit conflict about taking water during someone else's turn (*wārī/nōbat*). Water is available all year round in most hydraulic units such as in Hunza. However, where water is mainly available only in the spring, such as in Gilgit, water is distributed in specific ratios among clans or groups of water-rights holding families. Such rotation systems are socio-technical systems for water allocation, described as fixed turns (*wārābandī*) or protective irrigation, and are common not only in Gilgit-Baltistan but in the whole north-west of India and Pakistan. The *wārābandī* system is used to distribute water equally over large populations and areas under cultivation. It is not meant to match water supply with crop requirements, but to protect farmers against crop failure and famine (Narain 2008). User groups are allocated turns on the basis of neighbourhood, clan, or family, with usage measured in *ghāṛō* (around four inches in pipe diameter) and minutes (Grieser 2018). To get the water to their fields, the user has to close all irrelevant and open all relevant junctions of the canal and channel system, which is best done by those whose turn it is.

Depending on the availability of water, the flow has to be monitored in order to prevent the possible theft of water that can easily be proven with fields that are wet when it is not the owner's turn. The involvement of state-based institutions such as the police is rather uncommon. Instead, local arbitrators like a village representative (*nambardār*), councillor, or village elder (*jirga*) are involved (Grieser 2018). Traditionally, a perpetrator can be fined by the *nambardār* who would for example confiscate utensils or animals and hand them over to the offended household. However, with the introduction of Pakistani law which is different from customary laws, local arbitrators no longer have the legal right to confiscate and can be sued for such actions (Grieser 2018). Conflict could be over the rotation system, amount of water allocated, or access to water or waterways between different rights-holding groups (Grieser 2018).

Drinking water is traditionally extracted from water channels in the early morning and stored in neighbourhood or clan-specific earth pits (*gulko*). But with increasing populations, many without water rights, many places have to determine how these populations are to be supplied with drinking water. In Gilgit City, piped drinking water is commonly granted except during the spring when water is short. This means that during the spring season, public pipes may remain dry for specific areas that have been irrigated or settled only after 1947 and that are no longer settled by old settlers. Those without water rights are however allowed to collect water manually from the stream throughout the year with only the systematic water supply through the network being stopped (Grieser 2018). Those who hold water rights to spring water sources may be reluctant to share water, fearing a scarcity in irrigation water.

The perpetuation of traditional water rights has led to a debate on distributive fairness, with the idea of uniform needs of everyone, ultimately implying the relinquishing of customary rights (Grieser 2018). But while a modern understanding of the state usually aims at equality, the principle of equity or 'social justice' in Gilgit-Baltistan has been maintained, thereby continuing traditional water rights based on local historical norms (Grieser 2018). Land and water rights disputes thus 'often have historical legacies that span generations' with resources within the former princely kingdoms across Gilgit-Baltistan controlled by princes and prominent families (or more recently and in some cases, by the state), and where access is 'often determined by the power base' and 'protected by prominent groups' (AKF and AKPBS-P 2014: 11). Consequently, the idea of traditional rights itself has evolved into modern versions of tradition, with rights holders and the basis of these rights continuously amended (Grieser 2018; Boelens 1998, 2009; Trawick 2001; Wutich et al. 2015). 'Old settlers', however, keep thinking of the rules for *nallah* water distribution as rights (*haq*) in the form of a law (*qānūn*) and even as basic human rights (*buniyādi insāni huqūq*) (Grieser 2018).

Conflict over water is generally resolved by the *jirga, nambardār,* or local councillor, as resorting to the police or the courts is considered lengthy and humiliating for those involved and for the whole community, which would be seen as unable to resolve internal conflicts (Beg 2018). Conflicts that go to court can also end up dragging on for decades. If the conflict cannot be resolved at the village level, communities may opt for counselling and arbitration boards introduced by communities in some districts. Two approaches to conflict resolution can be identified in the WASEP model. The first is part of the social mobilization process and relies on building community capacities so that WSCs can resolve conflicts that may arise during project implementation (AKF and AKPBS-P 2014). Social mobilization is thus seen as key to mediating conflicts through continuous engagement (Datoo 2012). The precondition that 'any disputes associated with the ownership or utilisation of the water source or the land used for intake chambers, pipes or storage chambers … [are] resolved' before 'the commencement of any WASEP scheme' serves as an incentive for communities to tackle long-term disputes (AKF and AKPBS-P 2014: 12). Similarly, the requirement of community participation and support and to register demand for WASEP further reinforces the need to resolve any conflict before project implementation.

The second approach to conflict resolution and stabilization involves AKAH playing a lead role in identifying 'mechanisms to strengthen community organizations' and addressing underlying and wider social issues related to resource rights and equitable access to water (AKF and AKPBS-P 2014: 11–12).

As inter-communal and intra-communal tensions are related to the lack of access by marginalized groups within communities, AKAH identifies sources of perceived or actual inequities to ensure 'that all members of the targeted community have access to water and that individual needs are addressed' (AKF and AKPBS-P 2014: 12). These inequities could be between communal groups (within and between villages); between men and women, children, people living with disability, and the elderly; and among the 'ultra poor' who 'often have limited access and are disenfranchised [and] thus easily drawn to conflict' (AKF and AKPBS-P 2014: 12).

AKAH leads the mediation process with one of its teams negotiating 'a formal water sharing agreement' between the community/communities by 'engaging local religious leaders and influential people in a series of dialogues' (AKF and AKPBS-P 2014: 13). AKAH thus not only works to strengthen community capacity for communities to resolve conflicts as a precondition for project implementation, but also directly intervenes and takes the lead in resolving long-standing conflicts such as land and water disputes. The groundwork for AKAH's conflict resolution was laid by the Aga Khan Rural Support Programme (AKRSP) which established formalized community structures in many villages of Gilgit-Baltistan from the 1980s, locally referred to as *tanzīm* (organization) or village organizations, women's organizations and local support organizations, as long-term institutions (AKRSP 2011). While communal and collective work until then was often equated with the forced labour commissioned by the local rulers, the engagement of AKRSP transformed these into the notion of a gift for the community (Hussain and Langendijk 1995; Miller 2015). The WASEP model of community mobilization, contribution and participation, and conflict management, relies heavily on previous work by AKRSP. This also means that this model is known and appreciated mainly in those districts, villages, and communities in which AKRSP has previously worked.

7.3 Water-related conflict and conflict resolution in WASEP

This section examines and discusses the evidence on conflict based on the household survey, looking at the pattern of conflict across Gilgit-Baltistan by district, location (rural/urban), and community composition. It focuses on Jutial and Danyore to try to explain differences in perceptions of conflict reported by households in all seven Jutial and seven of the nine Danyore projects sampled. Information from WSC interviews and FGDs are used to explain differences in conflict between rural and urban projects, and the reasons for conflict in individual projects.

One of the key accomplishments of WASEP's social mobilization is said to be 'the successful resolution of community conflicts around the water rights and supply' with AKAH's social mobilization team 'enabling social cohesion and community stabilization through continuous dialogues, identification and resolution of conflicts, clarifying the interests and power sharing' (Datoo 2012: 11). This is supported by AKAH's own documentation and the household survey data. Of the 16 resolved conflicts out of 101 WASEP drinking water supply schemes (DWSS) in Gilgit-Baltistan and neighbouring Chitral in 2010–2014, 12 (75%) were over water rights and 3 (18.8%) about water source (AKF and AKPBS-P: 2014) (see Table 7.1). This illustrates the significance of water rights and access to water as sources of conflict in rural WASEP DWSS.

Evidence of conflict in WASEP from the household survey is drawn from five questions on unity, collective/harmonious water decisions, arguments over water, and fair allocation (Table 7.2). The results here only include positive and negative responses with neutral ('some', 'about the same as before') and other responses ('don't know'), along with non-responses ('N/A'), excluded; hence the total percentage for each question does not equal 100%. The row on 'community composition' indicates if the majority of projects in each district is homogeneous (HOM) or heterogeneous (HET) in terms of social composition. Homogeneous is defined as settlements where 90% or more of the population are from the same sectarian grouping. Heterogeneous is where less than 90% of the population are from the same sectarian grouping. This was based on information provided by AKAH on a village/settlement level as it was politically too sensitive to include this question in the household survey.

The household responses to all five conflict-related questions by district are positive overall, with Diamer recording the highest levels of village/area unity (100% of responses), collective water decisions (96.9%), harmonious water decisions (96.9%), no arguments over water (100%), and fairness of water allocation (100%) (Table 7.3). In contrast, Gilgit had the third lowest levels of village/area unity (58.7% of responses) and lowest positive responses for the four remaining questions: collective water decisions (45.5%), harmonious water decisions (40.3%), no arguments over water (83.7%), and fairness of water allocation (47.5%).

Given the higher number of urban projects concentrated in Gilgit, the differences between Diamer (100% rural) and Gilgit (96.4% urban) may be a reflection of differences between rural projects (1,125 household responses) and urban projects (2,007 household responses). Rural projects have consistently and noticeably more positive responses across all five conflict-related questions (Table 7.4). Similar differences can be observed in the positive responses when dividing all Gilgit-Baltistan projects into homogeneous (GB HOM) and heterogeneous (GB HET) groups. One reason for this difference is that urban populations are more diverse. For example, the average number of languages spoken in rural projects is one but is five in urban projects. Furthermore, 86% of rural respondent households lived in their settlement for over 50 years compared to only 49% of urban households. Conversely, 11% of urban respondents lived for less than five years in their current settlement while almost no rural households (0.005%) had less than five years' residence.

Table 7.1 WASEP resolved conflicts: KfW funding, Phases 1–4, rural DWSS, 2010–2014

Village	*AKAH mediator*	*Nature of conflict*	*Duration of mediation*	*Tools and medium*
Gulakhmuli	Social mobilization	Water rights	4 visits/meetings	Involved local council, local elders
Dodoshot	PRA	Water rights	2 hours	Formed committee, shared project design, involved external social capital
Barsat Teru Ghizer	Social/engineering	Water rights	6 months	Several meetings involving local council, local leaders, union council chairman
Pakora Ghizer	Community/WASEP	Water rights	15 years	WASEP involvement/community
Bakhtoli Chitral	Social mobilization	Water rights	3 years	WASEP involvement/community
Hatoon	Social mobilization	Water rights	12 years	Local council/community
Khanabad	Social mobilization	Water rights	12 years	Local council/community
Herkush	WASEP/community	Water rights	15 years	Local council/community
Helti	WASEP/community	Water rights	15 years	Local council/community
Guliakhturi	WASEP/community	Water rights	15 years	Local council/community
Immit/Nasirabad	WASEP/community	Water rights	1 year	Community
Burbur	WASEP/community	Water source	15 years	Community
Yashi Ghizer	WASEP/community	Water rights	10 years	Community
Junali Kuch	WASEP/community	Water source	5 years	Community
Singoor Dolomuch	WASEP/community	Source/feeder line	N/A	Community
Khudkusht	WASEP/community	Water source	N/A	Community

Note: KfW, German state-owned development bank; AKAH, Aga Khan Agency for Habitat; PRA, participatory rural appraisal; WASEP, Water and Sanitation Extension Programme.
Source: AKF and AKPBS-P 2014: 37–38.

Table 7.2 Household survey: social mobilization (SM) and conflict questions

Code	*Topic*	*Question*	*Answer options*
SM35	Village/area unity	In your opinion, how much unity is there in your village/area compared to other villages/areas that you know?	1 = Very little 2 = Not much 3 = Some 4 = Quite a lot 5 = A lot
SM36	Collective water decisions	In your opinion, are decisions about water in your village now made more collectively than before the WASEP project?	1 = Much less collectively 2 = Less collectively 3 = About the same as before 4 = More collectively 5 = Much more collectively
SM37	Harmonious water decisions	In your opinion, are decisions about water in your village now made more harmoniously than before the WASEP project?	1 = Much less harmoniously 2 = Less harmoniously 3 = About the same as before 4 = More harmoniously 5 = Much more harmoniously
SM38	Argument over WASEP water	Have you or a member of your household ever argued with someone in your area over WASEP water?	1 = Yes 2 = No
SM39	Fair WASEP water allocation	How fair do you feel the allocation of WASEP water among households is in your village or area?	1 = Very unfair 2 = Unfair 3 = Can't decide 4 = Fair 5 = Very fair

Table 7.3 Conflict indicators in WASEP sample schemes: Household responses, by district (%)

	GB	*Astore*	*Diamer*	*Ghizer*	*Gilgit*	*Hunza*	*Nagar*	*Shigar*	*Skardu*
Community composition:		*HOM*	*HOM*	*HET*	*Mix*	*HOM*	*HOM*	*HET*	*HOM*
SM35 Village/area unity									
A lot, quite a lot	69.6	85.0	100.0	77.4	58.7	78.0	58.6	61.7	57.5
Not much/very little	11.3	16.7	0.0	5.2	13.2	4.3	17.2	27.2	32.8
SM36 Collective water decisions									
More, much more collectively	54.2	71.7	96.9	71.3	45.5	59.6	63.8	50.6	52.2
Less/much less collectively	16.6	16.7	0.0	17.4	16.2	12.2	10.3	29.0	20.9
SM37 Harmonious water decisions									
More, much more harmoniously	50.9	58.3	96.9	66.6	40.3	56.9	72.4	61.1	56.0
Less/much less harmoniously	16.3	6.7	0.0	21.8	15.3	10.8	8.6	27.2	17.9
SM38 Arguments over water									
No	88.0	91.7	100.0	95.1	83.7	90.5	89.7	94.4	90.3
Yes	3.7	9.1	0.0	2.6	2.8	8.4	7.7	4.6	3.3
SM39 Fair water allocation									
Fair, very fair	61.6	70.0	100.0	83.2	47.5	76.2	75.9	67.3	76.1
Unfair/very unfair	14.0	23.3	0.0	9.8	14.4	13.0	12.1	25.3	10.4

Note: GB (Gilgit-Baltistan), HOM (homogeneous), HET (heterogeneous). Other responses or non-responses are not included, hence totals will not equal 100%. Danyore still under construction.

Table 7.4 Conflict in WASEP sample schemes, by community composition (%)

	Rural	*Urban*	*GB HOM*	*GB HET*	*Jutial HOM*	*Jutial HET*	*Danyore HET*
SM35 Village/area unity							
A lot, quite a lot	71.5	62.4	69.1	66.8	63.1	73.2	62.1
Not much/very little	8.4	11.9	8.1	11.0	5.7	3.8	14.5
SM36 Collective water decisions							
More, much more collectively	65.3	47.8	62.9	50.3	55.4	59.9	37.2
Less/much less collectively	17.5	16.0	15.2	13.0	10.8	8.3	10.0
SM37 Harmonious water decisions							
More, much more harmoniously	65.5	42.7	60.5	49.6	49.9	62.4	37.9
Less/much less harmoniously	18.5	10.1	16.2	12.6	11.0	5.7	9.7
SM38 Arguments over water							
No	93.9	84.8	94.4	84.5	94.9	99.4	72.4
Yes	3.5	3.2	2.8	3.3	1.8	0.0	1.7
SM39 Fair water allocation							
Fair, very fair	79.6	51.6	75.9	48.1	70.0	74.5	19.3
Unfair/very unfair	13.1	14.6	11.7	17.6	7.7	10.8	23.1

Note: GB (Gilgit-Baltistan), HOM (homogeneous), HET (heterogeneous). Other responses or non-responses are not included, hence totals will not equal 100%. Danyore still under construction.

Urban projects in Gilgit and specifically in Jutial and Danyore provide an opportunity to compare homogeneous and heterogeneous projects (Jutial) and heterogeneous projects (Danyore, 100% heterogeneous) (see Table 7.4). Paradoxically, heterogeneous Jutial projects (Jutial HET) have more positive responses across all five conflict-related questions than homogeneous Jutial (Jutial HOM) and Danyore (Danyore HET) projects. One possible explanation for this could be the more intense SM process in heterogeneous communities where there is a greater need for communities to make it work through more engagement, discussions, and by convincing more people. Very high numbers of households in Danyore also necessitated a top-down approach with community representatives making most decisions.

Disaggregating Jutial and Danyore projects is thus necessary to better understand the types of conflicts that affect WASEP urban projects and hence scalability. A breakdown of homogeneous and heterogeneous Jutial samples shows that Yasin Colony and partially Astore Colony (both homogeneous) recorded the lowest, and Noor Colony (heterogeneous) the highest, positive responses to conflict-related questions (Table 7.5).

Only seven out of the nine Danyore sample projects were included, with Hussainpura etc. and Princeabad Bala etc. omitted as the data was unreliable due to the high proportion of non-responses ('N/A') for a number of questions, possibly because Danyore projects were still under construction, having been held up by legal challenges (see below). Overall positive responses from the seven individual Danyore projects are significantly lower than Jutial projects and urban projects as a whole for several settlements and questions. These include very low positive responses for unity (Amphary Patti, Sharote etc.), collective water decisions (Amphary Patti, Hunza Patti, Shangote Patti, Sharote etc.), harmonious water decisions (Amphary Patti, Hunza Patti, Shangote Patti, Sharote etc., Sultanabad 1&2), and fair water allocation (Amphary Patti, Chikas Kote, Hunza Patti, Shangote Patti, Sharote etc.) (Table 7.6). Sharote etc. in particular records the most negative responses, with the lowest levels of village/area unity (11.4% of responses), collective water decisions (11.4%), harmonious water decisions (3.4%), and fairness of water allocation (4.5%).

One of the limitations of the household survey is that responses to the question on arguments over water are consistently positive and do not reflect responses to the other four questions, even when positive responses to these are low. To better understand where conflict has occurred, and the nature of water-related conflict, the household survey data is supplemented by interviews with WSCs and local social mobilizers, and FGDs. Conflict mainly occurred in the planning and implementation phases (Table 7.7) with the main reasons for both rural and urban conflict being concerns over resource distribution (including water shortages) (34.6%), objections to the payment of tariffs (18.9%), and the belief in (free) public provision (14.6%) (Table 7.7). In most cases, there is more than one reason given for conflict, hence the opposition to WASEP may have multiple reasons in most projects.

Table 7.5 Conflict in WASEP sample schemes: Jutial projects

	Aminabad	*Astore Colony*	*Diamer Colony*	*Noor Colony*	*Noorabad Extension*	*Wahdat Colony*	*Yasin Colony*	*Zulfiqarabad*
Community composition:	*HET*	*HOM*	*HOM*	*HET*	*HOM*	*HOM*	*HOM*	*HOM*
Settler status (%):	New (90.9)	New (92.6)	New (94.3)	New (88.8)	New (92.4)	New (85.4)	New (87.5)	New (79.3)
SM35 Village/area unity								
A lot, quite a lot	70.1	54.4	60.0	76.3	74.7	67.1	50.0	69.5
Not much/very little	2.6	5.9	7.1	5.0	3.8	11.0	1.4	4.9
SM36 Collective water decisions								
More, much more collectively	45.5	35.3	67.1	73.8	55.7	54.9	47.9	68.3
Less/much less collectively	9.1	5.9	11.4	7.5	1.3	23.2	16.7	6.1
SM37 Harmonious water decisions								
More, much more harmoniously	54.5	26.5	55.7	70.0	54.4	57.3	37.5	63.4
Less/much less harmoniously	3.9	5.9	11.4	7.5	5.1	17.1	20.8	6.1
SM38 Arguments over water								
No	98.7	98.5	95.7	100.0	98.7	91.5	86.1	98.8
Yes	0	1.5	0	0	1.3	4.9	1.4	1.2
SM39 Fair water allocation								
Fair, very fair	67.5	88.2	70.0	81.3	60.8	80.5	44.4	75.6
Unfair/very unfair	19.5	5.9	10	2.5	8.9	7.3	8.3	6.1

Note: Other responses or non-responses are not included, hence totals will not equal 100%. Community composition: HOM (homogeneous), HET (heterogeneous). Settler status: Old (90% and more have lived in settlements for at least 50 years), new (90% and more have lived in settlements for less than 50 years).

Table 7.6 Conflict in WASEP sample schemes: Danyore projects

	Amphary Patti (Mbad)	*Chikas Kote*	*Hunza Patti (Mbad)*	*Shangote Patti (Mbad)*	*Sharote etc.*	*Sultanabad 1&2*	*Syedabad*
Community composition:	*HET*	*HET*	*HET*	*HET*	*HET*	*HET*	*HET*
Settler status (%):	Old/new (50/50)	Old (97)	Old (80)	Old (70)	Old (N/A)	Old (N/A)	Old (80)
SM35 Village/area unity							
A lot, quite a lot	39.6	68.5	71.7	47.9	11.4	58.8	64.6
Not much/very little	15.1	7.9	3.3	19.8	54.5	20.2	19.0
SM36 Collective water decisions							
More, much more collectively	34.9	50.6	39.1	32.2	11.4	54.6	65.8
Less/much less collectively	15.1	12.4	3.3	19.0	65.9	24.4	24.1
SM37 Harmonious water decisions							
More, much more harmoniously	26.4	51.7	38.0	24.8	3.4	46.2	64.6
Less/much less harmoniously	14.2	11.2	0.0	22.3	73.9	26.9	25.3
SM38 Arguments over water							
No	67.0	97.8	77.2	84.3	88.6	91.6	88.6
Yes	1.9	0.0	3.3	6.6	4.5	5.9	3.8
SM39 Fair water allocation							
Fair, very fair	30.2	13.5	14.1	47.9	4.5	69.7	51.9
Unfair/very unfair	17.9	21.3	3.3	15.7	28.4	16.0	40.5

Note: Other responses or non-responses are not included, hence totals will not equal 100%. Community composition: HOM (homogeneous), HET (heterogeneous). Settler status: Old (90% and more have lived in settlements for at least 50 years), new (90% and more have lived in settlements for less than 50 years).

Table 7.7 Conflict: phase, reasons, mediation (%)

		Rural	*Urban*
Project phase	All	40.0	56.0
	Planning	66.7	90.9
	Implementation	55.6	54.5
	Post implementation	0.0	36.4
Reasons for conflict	Resource distribution/water shortage	20.0	14.6
	Water rights	10.0	0.0
	Extortion/free connections	0.0	9.8
	Prejudice against AKDN	6.7	2.4
	Lack of trust	3.3	2.4
	Lack of interest	6.7	0.0
	Lack of unity	6.7	0.0
	Tariffs	6.7	12.2
	Risks, natural hazards	3.3	2.4
	Public provision	0.0	14.6
	Pipe route	3.3	4.9
Mediators	Community members	8.7	23.3
	Local social mobilizers	8.7	11.6
	AKAH	13.0	11.6
	Individuals	4.3	0.0
	Umbrella organization	4.3	4.7
	WSCs	26.1	9.3
	Local leader	21.7	18.6
	Local notables	0.0	7.0
	Religious leader	13.0	0.0
	Government institution	0.0	9.3
	Courts	0.0	4.7
Conflict resolution methods	Dialogue	8.6	23.5
	Inclusion, accommodation	11.4	2.9
	Pipe realignment	14.3	2.9
	Religious references	14.3	20.6
	Health benefits	17.1	20.6
	Cultural references	14.3	17.6
	Litigation	0.0	2.9
	Coordination with public institution	14.3	2.9
	Information dissemination	5.7	5.9

Table 7.8 WASEP WSCs reporting conflict, rural

Settlement	*Rural/ urban*	*District*	*Social composition*	*Settler status: Under 5 yrs (>50 yrs) (%)*	*Reasons for conflict*
Broshal	Rural	Nagar	HOM	0 (88.9)	Prejudice against AKDN Pipe route
Chandupa	Rural	Skardu	HOM	1.5 (75.8)	Lack of interest Lack of unity Resource distribution Tariff payment
Daeen Chota	Rural	Ghizer	HOM	2.3 (88.6)	Prejudice against AKDN Lack of interest Lack of unity Tariff payment
Diruch	Rural	Ghizer	HOM	0 (95.7)	Resource distribution
Hasis Paeen	Rural	Ghizer	HOM	0 (84.3)	Resource distribution Water rights
Hatoon Paeen	Rural	Ghizer	HOM	0 (86.8)	Resource distribution Water rights
Hyderabad Center	Rural	Hunza	HOM	0 (96.1)	Resource distribution Water rights
Kuno	Rural	Ghizer	HOM	0 (94.9)	Lack of trust
Nasir Abad Ishkoman	Rural	Ghizer	HOM	2.1 (95.8)	Lack of unity Risks (natural hazards)
Shamaran Paeen	Rural	Ghizer	HOM	0 (86.7)	Resource distribution

Note: HOM, homogeneous; AKDN, Aga Khan Development Network

Conflict was reported in 10 (40%) rural and 14 (56%) urban projects from the sample (Tables 7.8 and 7.9). Conflict in rural projects related to resource distribution (20%), water rights (10%), and the lack of unity (10%) (Table 7.8). For urban projects, conflict was due to resource distribution (14.6%), the belief in (free) public provision (14.6%), tariff payments (12.2%), and extortion (9.8%) (Table 7.9). 'Extortion' includes individuals attempting to secure free connections by denying access to land (for pipe laying or water storage) or by withholding tariff payments. Conflict around 'tariff payment' appears to be related to the view that water services are a 'government responsibility' and should be free.

A crucial difference in the conflict in rural and urban projects is the duration of conflict – between 1 and 6 months (average 3 months) for rural projects compared to 3–24 months (average 9.5 months) for urban projects. Conflict mediation is conducted by a number of different parties, often in tandem rather than exclusively, with the main mediators being local leaders (40.3%), WSCs (35.4%), community members (32.0%), AKAH (24.7%), and local social

Table 7.9 WASEP WSCs reporting conflict, urban

Settlement	*Rural/ urban*	*District (Jutial/ Danyore)*	*Social composition*	*Settler status: Under 5 yrs >50 yrs (%)*	*Reasons for conflict*
Aliabad Centre	Urban	Hunza	HOM	0 82.6	Resource distribution Extortion Prejudice against AKDN Risks (natural hazards)
Amphary Patti	Urban		HET	8.5 65.1	Resource distribution Extortion
Astore Colony	Urban	Gilgit (J)	HOM	41.2 1.5	Lack of trust Public provision
Chikas Kot	Urban	Gilgit (D)	HET	1.1 93.3	Resource distribution Extortion Public provision
Chokoporo Gahkuch Bala	Urban	Ghizer	HOM	0 90.0	Public provision
Domial Buridur Gahkuch Bala	Urban	Ghizer	HOM	2.5 97.5	Public provision
Hunza Patti (Mbad)	Urban	Gilgit (D)	HET	2.2 41.3	Resource distribution
Hussainpura etc.	Urban	Gilgit (D)	HET	11.8 58.8	Tariff payment
Khanabad	Urban	Ghizer	HOM	0 91.4	Public provision
Rahimabad Aliabad	Urban	Hunza	HET	3.9 81.8	Pipe route
Sakarkoi	Urban	Gilgit	HET	19.3 7.0	Resource distribution Pipe route
Shangote Patti Mbad	Urban	Gilgit (D)	HET	5.0 59.5	Resource distribution Public provision
Sharote etc.	Urban	Gilgit (D)	HET	9.1 73.9	Tariff payment
Sultanabad 1&2	Urban	Gilgit (D)	HET	6.7 72.3	Tariff payment
Syedabad	Urban	Gilgit (D)	HET	5.1 53.2	Tariff payment
Wahdat Colony	Urban	Gilgit (J)	HOM	17.1 12.2	Tariff payment

Note: HOM, homogeneous; HET, heterogeneous

mobilizers (20.3%) (Table 7.7). Rural mediation relied mainly on WSCs (26.1%), local leaders (21.7%), AKAH (13%), and religious leaders (13%). Urban mediation was mainly conducted by community members (23.3%), local leaders (18.6%), local social mobilizers (11.6%), and AKAH (11.6%). The main tools of mediation relied on arguments about health benefits (37.7%), religious references (34.9%), dialogue (32.1%), and cultural references (31.9%), with health arguments, religious references, and cultural references utilized heavily in both but more so in urban projects. Rural projects also relied on pipe realignment (14.3%), and coordination with public institutions (14.3%), and the inclusion (accommodation) of opponents (11.4%), where urban projects relied most heavily on dialogue (32.1%).

Case study: Conflict in Danyore

As the largest urban WASEP project with over 9,000 households, Danyore serves as an important case study (see Chapter 2 for details), helping to illustrate the challenges in conflict resolution in the context of transferring and scaling up the WASEP model from rural to urban settings. Free utilities have been the norm in Gilgit-Baltistan for a variety of reasons, so even though not all households have access to clean drinking water, they are all aware that government schemes that provide water are free of charge. During the 2020 elections, nine different candidates announced free water for the Danyore area. In such a political landscape, demanding household investment in exchange for water provision can, in and of itself, trigger conflicts.

In instances like this, AKAH's team of social mobilisers has had to work extensively with communities to clarify the difference between the free schemes promised by politicians and what the WASEP model offers. This has been easier to accomplish in communities in which employees of AKAH have long-standing personal relationships, including their own families. This helps explain the pattern of site implementation across Gilgit-Baltistan, where earlier (rural) WASEP schemes benefited from community support through the work of AKDN agencies, including building local institutions and community capacities. These earlier sites have been described as low-lying fruit that served to demonstrate the successful implementation of WASEP to other communities.

Managing conflict is necessarily more complicated in peri-urban sites such as Danyore near Gilgit City, with more recent, diverse, and transient populations. Danyore has long-standing water allocation agreements between different parts of the community. The influx of households without multiple personal and familial relationships, combined with the significantly larger number of households requiring water, make these negotiations both more delicate and more relevant in the context of conflict management and scaling up. The households with a longer history of settlement wanted to implement an allocation pattern based on existing agricultural irrigation. Predictably, the residents of adjacent Muhammadabad and Sultanabad that are part of the Danyore scheme and who have historically been allocated less water for agriculture, wanted water to be distributed based on household numbers.

Unusually, AKAH began work on this WASEP project without having resolved the allocation of water to different *mohallah* (neighbourhoods) in

Table 7.10 Water distribution: Danyore scheme (%)

Water distribution mechanism	*Danyore*	*Sultanabad*	*Muhammadabad*
(a) Proposal by AKAH based on actual distribution of 9,050 households	60	17	23
(b) Proposal by Danyore leaders based on traditional water rights	80	10	10
(c) Agreed water distribution	75	12.5	12.5

Danyore. AKAH staff believed that provincial government intervention could have quickly resolved disagreements over the distribution formula of water in Danyore but was persuaded to rely on the *nambardārs* to work out an acceptable water distribution agreement with support of a provincial government representative. However, even if the state had the capacity to mediate, it was pointed out by the *nambardār* of Sultanabad that land and water disputes between Sultanabad and Danyore had been ongoing since 1928 without resolution. The contrast between water distribution based on (a) population and (b) historical water rights is illustrated in Table 7.10. After considerable negotiation with community notables from all three settlements, AKAH staff, and the district commissioner, AKAH produced a compromise solution (c) that allowed the project to progress. The conflicting interests in the case were and remain serious. Danyore residents were being asked to relinquish a part of their historical water allocation, while the residents of Sultanabad and Muhammadabad were being asked to accept a smaller allocation than the current population distribution might suggest. Such conflicts of interest are not insurmountable but must be built into any CBWM scheme since water rights from one time period are rarely, if ever, likely to be fit for a different time and population.

At the same time, a legal challenge was filed against the partnership between the Government of Gilgit-Baltistan (GoGB), AKAH, and communities for the 14 WASEP projects in Danyore by 11 opposition party candidates for the Gilgit-Baltistan Assembly, who argued that there was no provision in the Public-Private Partnership Act in Gilgit-Baltistan for partnerships between government and communities and that the agreement was invalid and community contributions unlawful because it was the government's responsibility to provide free drinking water. This legal action significantly delayed the completion of Danyore projects as some households chose not to contribute pending the outcome of the court case or because of the belief that water should be publicly provided and free.

In response, WSC members, volunteers, and AKAH staff visited households to explain how the scheme would operate, and AKAH proceeded with implementation, sharing progress regularly in the local media. With some households still not contributing, those in favour of the project decided to announce that they had won the (pending) court case and that anyone who paid immediately would get clean water or would be otherwise excluded.

In 2022 the Appellate Court ruled that the partnership between GoGB, AKAH, and the communities was binding. This case illustrates the potential politicization of a project and the impact that included a one-year delay and non-payment by households. Project delays have additional implications due to changes in costs, government leadership, and state personnel on the ground, all of which impact delivery.

7.4 Conclusion

The WASEP case illustrates how communities are not always homogeneous nor harmonious and community managed water projects will often face opposition that will require conflict management at different phases of a project. It is worth remembering that the selection of WASEP sites and requirements outlined in the terms of partnership (see Chapter 4), including community support of the project, seeks to minimize the risk of conflict and failure, thereby ensuring that settlements with unresolved conflict are excluded. Despite this, conflict still occurred in almost half of the WASEP sample. Given that urban projects are much more heterogeneous in terms of sectarian composition and the share of new settlers (in particular those arriving fewer than five years ago), urban projects can be expected to have lower participation rates (as seen in Chapter 5) and higher rates of conflict. In the case of WASEP, conflict was reported in 40% of rural and 56% of urban projects. Overall, conflict occurred mainly in the planning phase but was more frequent, lasted longer, and took longer to resolve in urban projects. Lower participation levels in urban households are related to the size of the project (total households), with participation negatively correlated with project size. Cooperation and participation are generally higher in rural households and households with higher levels of education, more members earning, and that have settled longer.

The nature of conflict also differs slightly, with rural communities concerned mainly about resource distribution (including water shortage) and, related to this, water rights. This is consistent with earlier AKAH documentation of conflict, if much less prominent (Table 7.1). For urban WASEP projects, conflict similarly centred on resource distribution but also arguments for (free) public provision (14.6%) and opposition to tariff payments (12.2%) that was likely related to political factors as described in the Danyore case study. The belief that water should be publicly provided and the occurrence of extortion are specifically urban phenomena with no cases reported in rural projects and is likely connected with political actors seeking political mileage. Curiously, given higher rural levels of homogeneity and participation, conflict due to the lack of interest (6.7%) and lack of unity (10.0%) are reported only in rural projects.

Overall, the main sources of conflict for rural and urban sites are over resource distribution (34.6%) and tariff payments (18.9%). These conflicts, in particular over tariff payments, need to be managed and resolved to ensure the sustainability and scalability of WASEP to urban areas, especially as political opportunism, where politicians promise free water to secure votes, can fuel opposition, undermine support for community contributions, and delay project implementation as was the case in Danyore.

Conflict mediation has mainly been led by local leaders, with AKAH to a lesser degree, in both rural and urban projects. WSCs and religious leaders were the other key mediators in rural projects compared to (individual) community members and local social mobilizers in urban projects. External mediators included AKAH and government institutions, with the courts being a last resort in the case of Danyore. The relatively small role played by urban WSCs suggests a possible lack of capacity to manage conflict especially as no training is provided for this. In the case of Jutial, where the majority of inhabitants are migrants from across Gilgit-Baltistan who do not have water rights to Jutial Nallah and Kargah Nallah, lifting water from Jutial Nallah through riverbank filtration was determined to be the best option to avoid conflict. The downsides are higher costs from operating a pump and the irregularity of water supply due to frequent (daily) power cuts.

CHAPTER 8
How financially sustainable is WASEP?

Jeff Tan, Sabrinisso Valdosh, and Saleem Uddin

8.1 Introduction

The Water and Sanitation Extension Programme (WASEP) is unusual in its successful delivery and continued operations of piped water networks as opposed to simple tap stands that typify much of community-based water management (CBWM). It is based on the principle of cost sharing in the CBWM model of community financing whereby the community contributes to the overall cost of the project or capital expenditure (CapEx) and through regular tariff payments for operating expenditure (OpEx), specifically operations and maintenance (O&M).

WASEP's sustainability depends on the regular collection of tariffs to cover costs and keep water systems operational. However, WASEP is constrained by a lack of willingness or ability of rural communities in particular to pay for water that is a feature of the CBWM model. The inherent tension between cost recovery and low tariffs, where irregular or non-payment of tariffs occur alongside very low tariff levels, undermines the financial sustainability of many rural water projects. Although urban WASEP projects fare better financially, sustainability also depends on whether communities can finance capital works associated with major repairs, and system rehabilitation and expansion.

This chapter examines the sustainability and scalability of WASEP in the context of these CBWM features. Section 8.2 expands on the features of CBWM that affect sustainability to provide a framework in which to assess financial and operational sustainability in WASEP. Section 8.3 provides an overview of how WASEP is financed in terms of CapEx. Section 8.4 examines how the operations of WASEP projects are maintained in terms of OpEx through day-to-day O&M related to minor repairs.

8.2 The constraints of community-based water management

The operational sustainability of WASEP is ultimately dependent on its financial sustainability, specifically the financing of CapEx and OpEx but also the lesser specified cost of capital maintenance expenditure (CapManEx) for major repairs and rehabilitation (Lockwood and Smits 2011). One key feature of CBWM is the ongoing reliance on external donor funding for water infrastructure because communities cannot and are not expected to cover the entire cost of CapEx or infrastructure. Nonetheless, community contributions to CapEx are meant to demonstrate demand and encourage community 'buy in' and hence 'ownership' of a water scheme. This is usually in the form of in-kind contributions of labour and local materials estimated at 15% to 66% of total finance for lower-middle-income countries (World Health Organization (WHO) and UN Water 2014) but is more typically 5–10% (9.5% in Pakistan) and often only a notional sum 'sometimes waived in order to speed up implementation processes' (Lockwood and Smits 2011: 113).

Unlike CapEx, communities are expected to cover the costs of OpEx or O&M 'through regular payment of tariffs or ad hoc contributions in cash and kind, if and when these costs arise' (Lockwood and Smits 2011: 114). However, CBWM is characterized by the irregular and non-payment of tariffs, and a tension between cost recovery and affordability; the very low tariff levels needed to ensure affordability undermine cost recovery and hence operational sustainability. There are widespread reports of irregular or non-payment of tariffs; the Pakistan Social and Living Standards Measurement Survey 2011–2012 estimated that only 22% of households (10% rural, 46% urban) pay for water at an average rate of PKR193 (US$1.97) per month (WHO and UN Water 2014). As a result, tariffs are often insufficient to cover O&M, especially for major repairs, and are 'seldom, if ever, expected to cover asset renewal or large-scale rehabilitation costs' associated with CapManEx (Lockwood and Smits 2011: 114).

This recurring theme in the literature (see e.g. Schouten and Moriarty 2003; Mugumya 2013; World Bank 2017; Hope et al. 2020) has implications for the operations and sustainability of water systems, with communities often having to 'wait for a major breakdown to occur and then fall back on local government, the NGO which implemented the original project, or donors to cover these much larger costs' (Lockwood and Smits 2011: 115). This is because there is often no clear differentiation between costs that are part of OpEx and those that are part of CapManEx which cover asset renewal, replacement, and rehabilitation, including 'work that goes beyond routine maintenance to repair and replace equipment in order to keep systems running' (Lockwood and Smits 2011: 26). The lack of a clear definition of, and distinction between, minor and major repairs, also creates confusion over what communities are expected to pay and when the government is obliged to step in (Lockwood and Smits 2011: 114).

Furthermore, funds for major repairs are not readily available and 'the agency or department approached by the community is not able to respond, leading to long breakdown times and failed services' (Lockwood and Smits 2011: 115). This is in part because CapManEx is 'the cost category that is least clearly understood, much less planned for in any systematic way' and 'meeting CapManEx is not even on the agenda in most countries, meaning that assets will continue to deteriorate' (Lockwood and Smiths 2011: 114, 116).

As a result, water services are not fully costed, whereas sustainable water services require full life-cycle costing not just for the lifespan of water infrastructure but indefinitely (Fonseca et al. 2011).

Two additional cost components are expenditure on direct support (ExpDS) to local level stakeholders, users or user groups for post-construction support activities, and expenditure on indirect support (ExpIDS) for 'macro-level support, [government] planning and policy making that contributes to the service environment' including 'developing and maintaining frameworks and institutional arrangements, and capacity building for professionals and technicians' (Lockwood and Smits 2011: 26, 117). However, very few countries 'specify the financing requirements for ... large-scale CapManEx and replacement expenditure and ExpDS and ExpIDS, including the vital function of post-construction support and monitoring' in part because '[p]olicy makers, governments and development partners of all descriptions have rarely, if ever, known the full costs of services' without which 'an informed debate about who should finance these costs is difficult if not impossible' (Lockwood and Smits 2011: 134, 25). The following sections examine how WASEP helps illustrate some of these issues.

8.3 WASEP financing: Capital expenditure

This section provides an overview of how WASEP is financed in terms of CapEx or the cost of physical infrastructure. WASEP is based on the CBWM model of financing where the bulk of total cost is covered by external sources, specifically donor funding that is channelled through government or directly to the implementing agency, in this case the Aga Khan Agency for Habitat (AKAH). As project implementation depends on donor funding, WASEP projects are implemented in phases, with each phase attributed to a (different) source of funding (Table 8.1).

The main funders of WASEP are almost all government or government-related agencies or bodies that receive public funds. KfW is the German state-owned development bank. The Pakistan Poverty Alleviation Fund (PPAF) is an autonomous not-for-profit company set up by Pakistan's Ministry of Finance in 1997, which receives funding from the International Fund for Agricultural Development, Agenzia Italiana per la Cooperazione allo Sviluppo (the Italian Agency for Development Cooperation), KfW, and the World Bank. Similarly, the Government of Gilgit-Baltistan (GoGB) received donor funding through the federal government for the most recent phase of WASEP in Jutial and Danyore, in Gilgit City.

The Japan Counter Value Fund is funded by Japan International Cooperation Agency. The European Commission and United States Agency for International Development are similarly state agencies while PATRIP Foundation was set up by KfW on behalf of the German government and funded by the German Federal Foreign Office, European Union, the governments of Denmark, Luxembourg, Norway, and Switzerland, and KfW. Only 7PV (Seven Priority

Table 8.1 WASEP financing: overview, funding phase

Funding phase	*Funding period*	*Total projects*	*Total households*	*Total population*	*Household taps*	*Community % average*	*Project share % average*	*Cost per person average*	*Cost per tap average*
KfW Phase I	1997–2001	67	7,277	62,876	7,433	37.8	62.2	PKR2,460 US$49	PKR20,808 US$414
Others	2001–2003	6	589	4,851	596	31.6	68.4	PKR2,770 US$46	PKR22,544 US$377
PPAF-I	2002–2004	11	593	5,398	606	32.7	67.3	PKR1,591 US$27	PKR14,168 US$242
PPAF-II	2004–2006	52	2,239	20,804	2,378	32.7	67.3	PKR3,142 US$53	PKR27,541 US$464
PPAF-V	2007–2008	50	1,742	19,648	1,900	32.8	67.2	PKR2,697 US$41	PKR28,486 US$434
ECD	2008–2010	3	404	3,636	463	34.2	65.8	PKR4,251 US$54	PKR21,663 US$274
PPAF-VI	2009–2009	14	448	4,648	562	40.3	59.7	PKR3,365 US$41	PKR31,298 US$383
PPAF-VII	2010–2012	26	1,830	16,707	1,919	36.5	63.5	PKR1,401 US$16	PKR12,032 US$136
EC	2010–2012	24	2,249	17,422	2,331	31.3	68.7	PKR5,265 US$60	PKR39,352 US$446
KfW Phase II	2010–2014	78	9,358	82,255	9,861	37.8	62.2	PKR2,957 US$32	PKR24,655 US$264
JCVF	2013–2016	72	6,220	55,298	6,128	44.6	55.4	PKR5,198 US$51	PKR46,538 US$455

(*Continued*)

Table 8.1 Continued

Funding phase	*Funding period*	*Total projects*	*Total households*	*Total population*	*Household taps*	*Community % average*	*Project share % average*	*Cost per person average*	*Cost per tap average*
USAID	2015–2017	1	485	3,880	513	39.6	60.4	PKR10,555 US$102	PKR79,829 US$772
7PV	2015–2018	23	1,565	13,316	1,578	41.2	58.8	PKR7,719 US$72	PKR65,265 US$605
GoGB*	2016–2021	25	12,390	99,120	12,436	16.5	83.5	PKR6,711 US$50	PKR53,492 US$399
PATRIP	2018–2018	7	240	2,132	241	44.8	55.2	PKR12,669 US$104	PKR111,631 US$918
Overall	**1997–2021**	**459**	**47,629**	**411,991**	**48,945**	**35.6**	**64.4**	**PKR4,296 US$53**	**PKR36,158 US$439**

Notes: *GoGB (Government of Gilgit-Baltistan) funding is officially recorded by AKAH as two urban projects (Jutial and Danyore) but each location has multiple schemes, each represented by its own WSC. PPAF, Pakistan Poverty Alleviation Fund; ECD, Early Childhood Development; EC, European Commission; JCVF, Japan Counter Value Fund. Costs in US$ are based on average exchange rates for each funding phase.

Valleys Programme) was funded internally by the Aga Khan Development Network and targeted two valleys in Gilgit-Baltistan (22 projects in Ghizer and two in Hunza) and three valleys in neighbouring Chitral (the largest district in Khyber Pakhtunkhwa province). No additional information was available for the Early Childhood Development (ECD) programme and 'others'.

Communities are expected to contribute to a share of total costs through in-kind contributions, mainly in the form of unskilled labour such as digging for pipe laying. In urban projects, communities tend to pay for the cost of this labour (based on the going daily wage rate and duration of the project), with only poor households opting to contribute labour. The exact share of community contributions is determined by AKAH during preparation of a bill of quantities and cost estimates (see Chapter 4) that are divided by the total number of households. The community and project (WASEP/donor) shares of contributions are calculated from WASEP multiple datasets that were merged and cleaned, with missing data estimated based on available information and similar percentages for each funding phase. As the cost per person depends on accurate estimates of household numbers and composition/size, it was necessary to stick with the original household numbers provided by AKAH that also informed the sample design even if subsequent field visits revealed higher household numbers due to population growth in several sample sites.

The average community contribution as a share of total project costs for WASEP is 35.6% (Table 8.1), which compares extremely favourably with 9.5% for CBWM schemes in Pakistan (WHO and UN Water 2014) and 5–10% for CBWM in general (Lockwood and Smits 2011). The community share is consistently high across nine districts in Gilgit-Baltistan, ranging from 33.6% (Shigar) to 55.2% (Diamer) with the exception of Gilgit district (19.6%) where 68% of WASEP projects are urban (Table 8.2). This is mirrored in the higher average rural community share (37.5%) compared to urban community contributions (23.7%) across Gilgit-Baltistan. This difference is most likely related to the higher cost of urban drinking water supply schemes (DWSS) (see Chapter 9); the average cost per person (PKR6,704 or US$56) and per household tap (PKR53,605 or US$451) for urban schemes is around 55% higher than in rural schemes in Pakistan rupees (Table 8.2). However, this is not reflected in the equivalent cost in US$ due to the depreciation of the Pakistani rupee from PKR41.1 to PKR163 (to the US dollar) between 1997 (at the start of WASEP) and 2021 (at the conclusion of this research). Based on an average annual exchange rate across the different funding phases, the overall US$50 cost per person of rural and urban piped water on premises for WASEP compares favourably to similar schemes in sub-Saharan Africa, lower-income and developing countries (Majuru et al. 2018; Torbaghan and Burrow 2019) (Table 8.3).

Higher levels of community contributions and lower costs for rural projects are also reflected in the WASEP sample of 25 rural and 25 urban projects (Tables 8.4 and 8.5). The average rural community share is slightly higher in the WASEP sample (46.4%) and cost per person (PKR5,620 or US$74) and per

Table 8.2 WASEP financing: district, rural, and urban

District	*Total projects*	*Average age in 2021*	*Total households*	*Total population*	*Household taps*	*Community % average*	*Project share % average*	*Cost per person average*	*Cost per tap average*
Astore	17	14.1	1,313	12,798	1,336	40.9	59.1	PKR2,739 US$40	PKR27,023 US$396
Diamer	6	6.2	402	5,402	432	55.2	44.8	PKR8,726 US$85	PKR115,667 US$1,131
Ghanche	25	9.9	1,883	17,154	1,820	48.2	51.8	PKR3,107 US$37	PKR28,749 US$348
Ghizer	188	11.3	13,620	119,809	14,260	40.4	59.6	PKR3,673 US$45	PKR30,998 US$380
Gilgit	50	6.0	16,129	131,758	16,305	19.6	80.4	PKR6,504 US$58	PKR52,434 US$471
Hunza	65	11.4	6,218	51,448	6,475	35.7	64.3	PKR4,502 US$57	PKR34,631 US$437
Kharmang	1	6.0	50	442	50	N/A	N/A	N/A	N/A
Nagar	15	13.5	1,413	12,554	1,424	39.2	60.8	PKR2,363 US$36	PKR20,123 US$306
Shigar	26	14.5	2,398	21,521	2,486	33.6	66.4	PKR3,521 US$52	PKR29,793 US$444
Skardu	66	12.7	4,203	39,537	4,361	42.8	57.2	PKR2,937 US$39	PKR26,129 US$349
Rural	**417**	**12.2**	**31,616**	**282,707**	**30,012**	**37.5**	**62.5**	**PKR3,565 US$46**	**PKR31,888 US$409**
Urban	**42**	**3.8**	**16,013**	**129,284**	**16,168**	**23.7**	**76.3**	**PKR6,704 US$56**	**PKR53,605 US$451**

Note: Costs in US$ are based on average exchange rates for each funding phase (see Table 8.1).

Table 8.3 Cost of water systems: selected countries/regions and WASEP

Sample	*Region*	*System description*	*Cost per person US$*
Angola	Urban	Piped on premises	283
Burkina Faso	Sahel Region/Mansila	Small piped scheme	131
Ghana	Ashanti	Piped water system	79
	Rural, small town	Piped water system	40
	Small towns	Piped water system	40–176
	Single town	Piped water network	36
	Multi-town	Piped water network	77
	Towns, villages	Reticulated supply	77
Mozambique	Urban	Piped on premises	24
	Single village, town	Piped water network	30–380
		Reticulated supply	87
		Urban household connections (full pressure)	725
South Africa		Piped water and house connection (full pressure)	410
Lower-income countries	Urban	High-cost technology water supply	319
	Urban	Intermediate technology water supply	159
	Urban	Peri-urban water supply	89–193
	Rural, peri-urban	Small piped system	47–130
		Medium piped system	30–267
		Peri-urban water	89–184
Developing countries		Piped on premises	218
Pakistan (WASEP)	**Gilgit-Baltistan Rural, peri-urban, urban**	**Piped on premises**	**50**

Note: WASEP costs in US$ are based on average exchange rates for each funding phase (see Table 8.1).
Source: Compiled from Majuru et al. (2018), Torbaghan and Burrow (2019), WASEP data.

tap (PKR48,247 or US$635) are significantly higher compared with the WASEP rural average of PKR3,565 (US$46) per person and PKR31,888 (US$409) per tap (Table 8.4). The urban share (21.3%) and costs (PKR6,483 per person or US$56, and PKR51,802 or US$442 per tap) for the WASEP sample are similar to the WASEP urban average (Table 8.5) and reflect the same pattern of rural–urban differences in community contributions and costs. Conversions to US dollars are based on average exchange rates for each funding phase unless otherwise specified and are included in the tables.

Table 8.4 WASEP financing: rural sample

Settlement	*District*	*Average age in 2021*	*Total households*	*Total population*	*Household taps*	*Community % average*	*Project share % average*	*Cost per person average*	*Cost per tap average*
Broshal	Nagar	23	81	652	89	34.4	65.6	PKR1,740 US$35	PKR12,748 US$254
Burdai	Skardu	18	34	272	35	29.4	70.6	PKR1,781 US$30	PKR13,837 US$236
Chandupa	Skardu	11	93	692	93	39.1	60.9	PKR4,854 US$97	PKR36,115 US$719
Daeen Chota	Ghizer	15	52	471	23	39.9	60.1	PKR2,430 US$41	PKR49,768 US$839
Diruch	Ghizer	20	75	600	79	42.0	58.0	PKR2,222 US$44	PKR16,876 US$336
Dushkin	Astore	15	133	1,200	72	32.8	67.2	PKR2,819 US$48	PKR46,990 US$792
Duskhore Hashupi	Skardu	13	20	241	24	40.7	59.3	PKR2,214 US$34	PKR22,237 US$339
Halpapa Astana	Skardu	9	87	714	95	29.4	70.6	PKR4,367 US$49	PKR32,821 US$372
Hasis Paeen	Ghizer	23	66	594	70	30.9	69.1	PKR2,569 US$51	PKR21,796 US$434
Hatoon Paeen	Ghizer	10	120	1,559	125	57.0	43.0	PKR12,049 US$129	PKR150,279 1,607
Hundur Barkulti	Ghizer	4	48	432	48	46.7	53.3	PKR6,627 US$61	PKR59,644 US$553

(Continued)

Table 8.4 Continued

Settlement	*District*	*Average age in 2021*	*Total households*	*Total population*	*Household taps*	*Community % average*	*Project share % average*	*Cost per person average*	*Cost per tap average*
Hyderabad Center	Hunza	9	100	760	150	31.0	69.0	PKR4,427 US$50	PKR22,430 US$254
Janabad	Hunza	7	75	626	75	36.0	64.0	PKR10,281 US$110	PKR85,808 US$918
Kirmin	Hunza	3	81	729	81	55.7	44.3	PKR8,362 US$69	PKR75,261 US$619
Kuno	Ghizer	8	71	660	75	61.9	38.1	PKR10,775 US$115	PKR94,821 US$1,014
Marikhi	Ghizer	9	186	1,488	220	38.8	61.2	PKR805 US$9	PKR5,446 US$62
Nasir Abad Ishkoman	Ghizer	10	53	477	55	52.7	47.3	PKR13,855 US$148	PKR120,161 US$1,285
Nazirabad	Ghizer	10	40	305	45	24.7	75.3	PKR9,835 US$105	PKR66,658 US$713
Rahimabad (Mdass)	Gilgit	9	120	896	131	44.4	55.6	PKR5,007 US$49	PKR34,249 US$335
Shamaran Paeen	Ghizer	13	24	288	24	39.1	60.9	PKR3,237 US$49	PKR38,848 US$592
Shilati	Diamer	8	46	582	54	52.9	47.1	PKR13,820 US$135	PKR148,953 US$1,456
Singul Shyodass	Ghizer	17	18	162	21	26.3	73.7	PKR777 US$13	PKR5,992 US$102

(Continued)

Table 8.4 Continued

Settlement	*District*	*Average age in 2021*	*Total households*	*Total population*	*Household taps*	*Community % average*	*Project share % average*	*Cost per person average*	*Cost per tap average*
Staq Paeen	Skardu	20	94	752	75	30.0	70.0	PKR3,196 US$64	PKR32,040 US$638
Sutopa	Skardu	13	40	480	46	32.7	67.3	PKR1,999 US$30	PKR20,860 US$318
Yuljuk	Skardu	11	61	404	63	42.5	57.5	PKR4,198 US$84	PKR26,919 US$536
Rural sample		**12.3**	**1,818**	**16,036**	**1,868**	**46.4**	**53.6**	**PKR5,620 US$74**	**PKR48,247 US$635**
Rural projects		**11.9**	**31,616**	**282,707**	**32,665**	**40.5**	**59.5**	**PKR3,565 US$46**	**PKR31,888 US$409**

Note: Costs in US$ are based on average exchange rates for each funding phase (see Table 8.1).

Table 8.5 WASEP financing: urban sample

Settlement	*District*	*Average age in 2021*	*Total households*	*Total population*	*Household taps*	*Community % average*	*Project share % average*	*Cost per person average*	*Cost per tap average*
Aliabad Centre	Hunza	7	360	2,762	413	29.0	71.0	PKR3,869 US$41	PKR25,877 US$277
Aminabad	Gilgit	8	300	2,800	320	16.4	83.6	PKR4,117 US$44	PKR36,027 US$385
Amphary Patti (Mbad)	Gilgit	0	850	6,800	850	19.5	80.5	PKR5,133 US$38	PKR41,061 US$306
Astore Colony	Gilgit	3	61	488	85	18.2	81.8	PKR13,764 US$103	PKR79,020 US$590
Chikas Kote	Gilgit	0	691	5,528	691	19.5	80.5	PKR5,133 US$38	PKR41,061 US$306
Chokoporo Gahkuch Bala	Ghizer	6	170	1,246	179	34.6	65.4	PKR9,288 US$91	PKR64,656 US$632
Diamer Colony	Gilgit	3	288	2,304	310	11.8	88.2	PKR12,764 US$95	PKR94,863 US$708
Domial Buridur Gahkuch Bala	Ghizer	8	59	371	59	42.1	57.9	PKR8,332 US$81	PKR52,392 US$512
Hassan Abad Aliabad	Hunza	10	77	693	82	55.8	44.2	PKR11,839 US$127	PKR100,053 US$1,070
Hunza Patti (Mbad)	Gilgit	0	551	4,408	551	19.5	80.5	PKR5,133 US$38	PKR41,061 US$306
Hussainpura etc.	Gilgit	0	1,620	12,960	1,620	19.5	80.5	PKR5,133 US$38	PKR41,061 US$306
Khanabad Gahkuch Bala	Ghizer	6	55	421	56	47.7	52.3	PKR6,821 US$67	PKR51,282 US$501
Noor Colony Urban Project	Gilgit	9	325	3,541	331	13.5	86.5	PKR6,418 US$69	PKR68,656 US$734
Noorabad Extension	Gilgit	7	325	2,600	325	17.8	82.2	PKR7,651 US$82	PKR61,210 US$655

(Continued)

Table 8.5 Continued

Settlement	*District*	*Average age in 2021*	*Total households*	*Total population*	*Household taps*	*Community % average*	*Project share % average*	*Cost per person average*	*Cost per tap average*
Princeabad Bala etc.	Gilgit	0	992	7,936	992	19.5	80.5	PKR5,133 US$38	PKR41,061 US$306
Rahimabad Aliabad	Hunza	9	321	2,204	321	26.9	73.1	PKR4,840 US$52	PKR33,234 US$355
Sakarkoi	Gilgit	9	116	1,047	120	55.8	44.2	PKR10,818 US$116	PKR94,385 US$1,010
Shangote Patti (Mbad)	Gilgit	0	590	4,720	590	19.5	80.5	PKR5,133 US$38	PKR41,061 US$306
Sharote etc.	Gilgit	0	630	5,040	630	19.5	80.5	PKR5,133 US$38	PKR41,061 US$306
Soni Kot	Gilgit	10	310	2,790	321	59.0	41.0	PKR9,497 US$102	PKR82,541 US$883
Sultanabad 1&2	Gilgit	0	799	6,392	799	19.5	80.5	PKR5,133 US$38	PKR41,061 US$306
Syedabad	Gilgit	0	292	2,336	292	19.5	80.5	PKR5,133 US$38	PKR41,061 US$306
Wahdat Colony	Gilgit	3	372	2,976	372	11.6	88.4	PKR12,915 US$96	PKR103,321 US$771
Yasin Colony	Gilgit	3	170	1,360	170	11.8	88.2	PKR12,914 US$96	PKR103,311 US$771
Zulfiqarabad	Gilgit	3	469	3,752	469	11.8	88.2	PKR12,764 US$95	PKR102,109 US$762
Urban sample		**6.5**	**10,793**	**87,475**	**10,948**	**21.3**	**78.7**	**PKR6,483 US$56**	**PKR51,802 US$442**
Urban projects		**3.8**	**16,013**	**129,284**	**16,168**	**20.1**	**79.9**	**PKR6,704 US$56**	**PKR53,605 US$451**

Note: Average age of urban samples excludes nine schemes in Danyore that were still under construction and only completed in 2022. Costs in US$ are based on average exchange rates for each funding phase (see Table 8.1).

8.4 WASEP financing: Operating expenditure

OpEx covers the day-to-day O&M related to fixing minor repairs and the cost of recurrent expenses such as energy and other bills and salaries. O&M is thus a key component of (operational and financial) sustainability. In the case of WASEP, sustainability is based on two main sources of revenue. The first is an upfront payment of PKR3,000 (rural) and PKR8,000 (urban) as connection fees, which is deposited into an O&M fund. (This roughly converts to US$18–73 (rural) and US$49–195 (urban) depending on the annual average dollar exchange rate of PKR41.1 at the start of WASEP in 1997, and PKR163 at the conclusion of the research in 2021.) This (endowment) fund is deposited in a bank as a term or fixed deposit (referred to as Term Deposit Receipt or TDR) with interest (or 'profit') rates of 5–14%.

The second is a monthly tariff that the community sets, which ranges from PKR100 (US$0.61–2.43) to PKR500 (US$3.07–12.16) per household, depending on socioeconomic condition and location (according to AKAH staff) although AKAH reports also mention tariffs of PKR200 (US$1.23–4.86) to PKR300 (US$1.84–7.30) (AKPBS-P and USAID 2016: 8). This tariff collection is meant to be deposited into a savings account along with profit transferred from the TDR. The profit on the TDR plus monthly tariffs are then used for O&M, minor repairs, and to pay the salaries of employees, most notably a water and sanitation officer (WSO) or plumber, and (female) water and sanitation implementer (WSI) for health and hygiene awareness, which is a key component of WASEP's integrated approach (see Chapters 4 and 6). AKAH believes that most water and sanitation committees (WSCs) have significant savings after paying salaries and other expenses, along with new connection fees that are also deposited in the savings account. Additionally, fines for the misuse of (drinking) water (e.g. for crops) and non-payment is seen as another source of revenue.

These revenue sources are expected to cover the two main expenditure items: maintenance and repairs. Maintenance involves repairs but is also distinct in that it is pre-emptive (e.g. line flushing and exercising valves) (AKPBS-P n.d.) to prevent or reduce future breakdowns and hence the disruption and cost of repairs, whereas repairs are generally reactive. A WSO can thus be considered central to operational sustainability, paid for through regular (monthly) tariffs as specified under WASEP's terms of partnership (see Chapter 4). Maintenance will often involve minor repairs in terms of fixing leaks and replacing parts such as valves. The cost of these minor repairs is meant to be covered by profits transferred from the O&M fund. AKAH estimates the average annual running cost per household for gravity-fed water systems at PKR60,000 (or US$379 based on annual exchange rates for 2019–2021) compared to PKR110,000 (US$695) for mechanized water systems. The operational and financial sustainability of WASEP thus depends on the amount of profit generated by the O&M fund, tariff level, and frequency of tariff collection and payments.

WSC revenue from the WASEP sample is calculated from monthly profit from the O&M fund plus tariff revenue. Where tariffs are not collected regularly (or are only collected when repairs are needed) then this is recorded as '0' under 'tariff collection'. WSC expenditure is calculated from estimates of monthly operating expenses, salaries, and minor repairs. Data on revenues and expenditure for the sample of 25 rural and 25 urban WSCs come from: (a) AKAH datasets for O&M funds (connection fee multiplied by total households); (b) interviews with WSC treasurers for interest rates for O&M funds, tariff collection (tariff multiplied by total connected and paying households), salaries, operating costs, and minor and major repairs; and (c) bank statements for O&M fund profits and current balance.

Revenue from new connections was not included because this data was not consistently available and would have been (partly) spent on the cost of installing new connections, and interest earned was captured in bank statements. As such, discrepancies in these estimates are unlikely to significantly alter the final operating results and overall findings in most instances. Moreover, expenses are likely to be underestimates given that where no amount is provided for costs associated with 'salaries' or 'minor repairs', these are entered as '0' (zero) in order to calculate estimates of expenditure. Finally, not all WSCs provided bank statements and 10 rural and 4 urban (joint) WSCs covered more than one project and shared the same bank account. Data for individual projects with joint WSCs and shared bank accounts with other WASEP projects that may or may not be part of the sample (Table 8.6) were disaggregated but it was not possible to completely isolate individual financial performance.

Table 8.6 WASEP joint WSCs and shared bank accounts, sample

Rural/ urban	*Sample**	*WASEP projects with joint WSCs and bank accounts*	*Households (total)*	*Households (%)*
Rural	Daeen Chota*	Daeen Bada	64	55.2
		Daeen Chota*	52	44.8
Rural	Hatoon Paeen*	Hatoon Dass	90	25.7
		Hatoon Omch	150	42.9
		Hatoon Paeen*	120	34.3
Rural	Hundur Barkulti*	Barkolti Bala	100	31.4
		Barkolti Paeen	170	53.5
		Hundur Barkulti*	48	15.1
Rural	Hyderabad Center*	Hyderabad Center*	87	18.8
		Barbar/Sheraz	41	8.8
		Chumarkhand	83	17.9
		Shahabad	100	21.6
		Brongshal	104	22.4
		Khrukushal	49	10.6
Rural	Janabad*	Janabad*	75	37.3
		Passu	126	62.7
Rural	Kuno*	Askamdass Cheerat	85	41.5

(Continued)

Table 8.6 Continued

Rural/ urban	*Sample**	*WASEP projects with joint WSCs and bank accounts*	*Households (total)*	*Households (%)*
		Karimabad Thoi	45	22.0
		Kuno*	75	36.6
Rural	Marikhi*	Gich	151	44.8
		Marikhi*	186	55.2
Rural	Shamaran Paeen*	Shamaran Bala	25	21.0
		Shamaran Centre	16	13.4
		Shamaran Muldeh	16	13.4
		Shamaran Paeen*	24	20.2
		Shamaran Tordeh	38	31.9
Rural	Singul Shyodass*	Singul Bala	75	27.8
		Singul Hunphari	34	12.6
		Singul Kayphari	59	21.9
		Singul Mujahid Muhalla	17	6.3
		Singul Paeen	67	24.8
		Singul Shyodass*	18	6.7
Rural	Sutopa*	Dass Khore	19	22.6
		Kachay	25	29.8
		Sutopa*	40	47.6
Urban	Aliabad Centre*	Rahimabad Aliabad*	321	31.7
		Aliabad Centre*	360	35.5
		Aliabad Sultanabad	333	32.8
Urban	Chokoporo Gahkuch Bala*	Chokoporo Gahkuch Bala*	170	48.7
		Domial Buridur Gahkuch Bala*	59	16.9
		Khanabad Gahkuch Bala*	55	15.8
		Pari Hinal Gahkuch Bala	65	18.6

Of the 25 rural WSCs, three (Kirmin, Shilati, Staq Paeen) had not deposited their O&M funds in bank accounts so there was no revenue from interest for maintenance or repairs. Eight (32%) WSCs did not collect tariffs regularly or at all, with five (Broshal, Burdai, Halpapa Astana, Kirmin, Shilati) collecting payments as and when needed for repairs, and one (Chandupa) only functioning for 20% of households (Table 8.7). As mentioned, tariff collection for these WSCs is recorded as '0' as collection on a needs basis centres on repairs as opposed to regular maintenance and is recorded under OpEx and minor repairs. Low levels of regular tariff collection are reflected in 12 (48%) WSCs not paying salaries, and hence not employing a (full-time) WSO. Of these, three WSCs (Broshal, Singul Shyodass, Sutopa) had managed to secure the services of a government plumber, paid for by GoGB, and three (Kirmin, Shilati, Yuljuk) relied on volunteers, negating the perceived need for monthly

Table 8.7 WSC average monthly revenue and expenditure: estimates, rural sample (PKR)

WSC	*O&M fund start*	*O&M fund interest*	*Tariff collection*	*Total revenue*	*OpEx*	*Salaries*	*Minor repairs*	*Total exp'd*	*Surplus (deficit)*	*O&M fund Aug 2021*	*O&M fund +/(–)*
Broshal	70,000	467	0	467	1,000	0	1,000	2,000	(1,533)	120,000	50,000
Burdai	34,000	227	0	227	1,000	0	1,000	2,000	(1,773)	30,000	(4,000)
Chandupa	104,400	829	0	829	N/A	0	N/A	N/A	N/A	203,637	99,237
Daeen Chota	160,000	1,333	750	1,750	5,000	3,000	2,917	10,917	(8,833)	245,504	85,504
Diruch	105,000	653	1,050	1,703	4,700	3,000	1,000	8,700	(5,433)	129,000	24,000
Dushkin	332,500	2,217	1,050	3,267	4,700	3,000	1,000	8,700	(5,433)	129,000	(203,500)
Duskhore Hashupi	60,000	250	1,365	1,615	4,500	3,000	3,333	10,833	(9,218)	20,000	(40,000)
Halpapa Astana	231,000	1,155	0	1,155	3,500	3,000	5,000	11,500	(10,345)	3,000	(228,000)
Hasis Paeen	70,000	428	2,457	2,885	7,500	3,000	833	11,333	(8,468)	210,000	140,000
Hatoon Paeen	597,000	3,163	4,900	8,063	1,000	4,000	83	5,083	3,797	150,000	(447,000)
Hundur Barkulti	144,000	848	1,550	2,408	1,560	0	4,167	2,810	(170)	132,276	(11,724)
Hyderabad Center	285,000	1,774	15,000	16,774	4,000	8,000	4,000	16,000	774	314,497	29,497
Janabad	225,000	2,255	7,197	5,885	2,964	3,000	2,000	7,964	(2,079)	95,861	(129,139)
Kirmin	243,000	No data	C	0	3,000	0	1,667	4,667	(4,667)	202,500	(40,500)
Kuno	213,000	2,441	1,975	4,416	1,200	2,000	1,167	4,367	49	101,016	(111,984)
Marikhi	108,000	887	3,674	4,561	1,000	0	1,000	2,000	2,561	119,784	11,784
Nasir Abad Ishkoman	159,000	1,195	1,800	2,995	0	0	250	250	2,745	150,000	(9,000)
Nazirabad	120,000	846	1,200	2,046	2,500	2,500	208	5,208	(3,163)	38,102	(81,898)
Rahimabad (M'dass)	360,000	1,372	4,200	5,572	2,000	4,000	2,083	8,083	(2,512)	350,000	(10,000)
Shamaran Paeen	60,000	514	765	1,279	4,000	2,200	4,000	10,200	(8,921)	32,724	(27,276)
Shilati	138,000	No data	0	0	1,500	0	0	1,500	(1,500)	200,000	62,000
Singul Shyodass	18,000	55	0	55	200	0	334	534	(479)	7,705	(10,295)
Staq Paeen	94,800	No data	3,680	3,680	3,600	0	4,000	7,600	(3,920)	15,000	(79,800)
Sutopa	120,000	800	1,638	2,438	1,000	0	833	1,833	605	N/A	N/A
Yuljuk	78,000	520	0	520	500	0	1,250	1,750	(1,230)	N/A	N/A

tariffs. The remaining rural WSCs resorted to household contributions as and when repairs were needed.

Given the low levels of tariff collection, only six (24%) rural WSCs are estimated to have a monthly operating surplus as at August 2021: Hatoon Paeen (PKR3,797), Hyderabad Center (PKR774), Kuno (PKR49), Marikhi (PKR2,561), Nasir Abad Ishkoman (PKR2,745), and Sutopa (PKR605). Similarly, of the 23 operational rural WSCs with data for the O&M fund (excluding Chandupa), eight (36.4%) recorded increases in the value of O&M funds, with the remaining 15 (65.2%) recording a lower balance than at the start of the project. This indicates O&M funds are being drawn down to pay for maintenance and repairs where tariff collection is insufficient, and may not be a sustainable revenue source. Operating deficits were partly mirrored in WSC responses where only nine (37.5%) WSCs reported being able to cover both OpEx and the costs of minor repairs and even then, five of these WSCs were estimated here to run monthly operating deficits (in brackets in Table 8.7): Hundur Barkulti (PKR170), Janabad (PKR2,079), Nazirabad (PKR3,163), Rahimabad (PKR2,512), and Singul Shyodass (PKR479).

The sustained operations of 24 out of 25 rural WSCs suggests that OpEx and minor repairs were being covered by irregular household contributions as and when needed. Similarly, of the 21 rural WSCs that required major repairs, only Hyderabad Center and Singul Shyodass paid for this from the O&M fund, with nine WSCs relying on additional community contributions (Broshal, Daeen Chota, Hatoon Paeen, Janabad, Kirmin, Kuno, Marikhi, Shamaran Paeen, Shilati). In Kuno, building an additional spring water chamber (PKR100,000 or US$878 based on average annual exchange rates for 2014–2019) was paid for with a personal donation from a village notable. Four WSCs also received additional (external) support: Daeen Chota and Hundur Barkulti (replacement pipes from AKAH), Halpapa Astana (grant from the Local Government and Rural Development Department), and Shamaran Paeen (unspecified external support).

Financial data for urban WSCs covered 16 projects as nine Danyore projects (Amphary Patti, Chikas Kote, Hunza Patti, Hussainpura etc., Princeabad Bala etc., Shangote Patti, Sharote etc., Sultanabad 1&2, Syedabad) were still under construction. Unlike rural WSCs, all 16 operational urban projects had their O&M funds deposited in bank accounts with overall higher profits from interest payments (Table 8.8). Almost all urban WSCs (81%) collected tariffs with only Chokoporo, Domial Buridur, and Khanabad not doing so having secured the services of a government plumber (paid for by GoGB). These communities were said to have influential members who requested and secured the services of a government plumber who was transferred from another non-functional government project in the same area. Six urban projects in total (37.5%) did not pay for the salary of a WSO. The range of the salaries reflects the number of staff employed and in turn the level of community organization. Unlike rural WSCs, average monthly expenditure was significantly higher, especially for Jutial projects (Astore Colony,

Table 8.8 WSC average monthly revenue and expenditure: estimates, urban sample (PKR)

WSC	*O&M fund start*	*O&M fund interest*	*Tariff collection*	*Total revenue*	*OpEx*	*Salaries*	*Minor repairs*	*Total exp'd*	*Surplus (deficit)*	*O&M fund Dec 2021*	*O&M fund +/(–)*
Aliabad Centre	1,080,000	12,620	66,327	79,767	4,022	144,170	0	88,222	(8,455)	4,295,403	3,215,403
Amin Abad	990,000	13,481	86,100	96,825	0	5,000	200,000	21,667	75,158	2,250,000	1,260,000
Astore Colony	488,000	3,253	25,500	28,753	15,000	5,000	5,000	25,000	3,753	1,260,000	772,000
Chokoporo Gahkuch Bala	510,000	13,120	0	13,120	867	0	40,000	2,863	10,257	657,450	147,450
Diamer Colony	2,304,000	15,360	93,000	108,360	16,571	32,689	15,000	64,259	44,101	2,400,000	96,000
Domial Buridur Gahkuch Bala	177,000	5,603	0	5,603	301	0	693	994	4,609	228,150	51,150
Hassan Abad Aliabad	231,000	1,230	2,310	3,540	2,000	0	40,000	42,000	(38,460)	219,000	(12,000)
Khanabad	165,000	5,226	0	5,226	281	0	647	927	4,299	213,300	48,300
Noor Colony	2,600,000	26,969	162,000	188,969	11,587	75,000	25,000	111,587	77,382	4,740,000	2,140,000
Noorabad Ext	2,600,000	17,333	121,500	138,833	40,000	0	2,083	47,083	91,750	2,000,000	(600,000)
Rahimabad Aliabad	963,000	12,620	84,750	78,947	4,022	69,036	6,888	79,946	(999)	3,835,613	2,872,613
Sakarkoi	348,000	4,642	32,000	36,642	20,000	0	12,500	42,500	(5,858)	700,000	352,000
Soni_Kot	930,000	13,372	130,000	143,372	100,000	100,000	8,333	208,333	(64,961)	100,000	(830,000)
Wahdat Colony	2,976,000	23,105	97,500	120,605	58,500	30,000	375	88,875	31,730	5,166,548	2,190,548
Yasin Colony	1,376,000	5,902	37,800	46,973	31,000	5,000	1,333	37,333	9,640	1,162,500	(213,500)
Zulfiqarabad	3,752,000	43,773	144,000	187,773	400,000	10,000	33,333	443,333	(255,560)	3,045,000	(707,000)

Diamer Colony, Noor Colony, Noorabad Extension, Wahdat Colony, Yasin Colony, Zulfiqarabad) where electricity bills for the mechanized water system accounted for the bulk of this. In contrast to rural WSCs, 10 (62.5%) urban WSCs are estimated to have operating surpluses, and 11 (68.8%) have increased the value of their O&M funds through increased connections and connection fees as a result of urbanization.

8.5 Household willingness to pay

Results from the household survey generally confirm the evidence from WSCs in the previous section. Based on questions around O&M (Table 8.9), only 35% of rural and urban households across Gilgit-Baltistan pay tariffs every month (O&M13); 89% of rural and 83% of urban households contributed in cash at the start to the O&M fund (O&M14) with a median payment of PKR3,000 (or US$38 based on average annual exchange rates for the year of completion across the sample) (rural) and PKR7,000 (US$67) (urban), similar to the stipulated connection fee (Table 8.10). As expected, the community share of project

Table 8.9 Household survey questions: operations and maintenance (O&M) and community contributions

Code	*Topic*	*Question*	*Answer options*
O&M13	Frequency of payment for WASEP water	How often does your household pay money for the project water?	At start only At start and every week At start and every month At start and every three months At start and every six months At start and once a year At start but then no fixed schedule Don't know
O&M14	Contribution at the start of WASEP	What did your household contribute at the start of the WASEP project?	Nothing Cash [PKR] In-kind contributions [describe] Don't know
O&M16	Labour contribution at start	How many hours of labour did your household contribute at the start of the WASEP project?	Number of hours Other Don't know Not applicable
O&M17	Payment for WASEP each month	How much money does your household pay for the WASEP project water each month?	Nothing Cash [PKR] In-kind contributions [describe] Don't know
O&M24	Cost of WASEP water	What is your opinion of the cost of WASEP project water in your village/area?	Very bad Bad OK Good Very good Don't know

costs through labour contributions (O&M16) comes mainly from rural (84%) compared to urban (29%) households; the median number of labour hours per household was eight. More crucially, there is a significant share (39%) of both rural and urban households that do not pay monthly tariffs (O&M17) as noted in the interviews with WSC treasurers; this figure was lower in homogeneous (28%) and higher in heterogeneous (48%) communities (Table 8.10).

The median amount of tariff paid (PKR33 or US$0.21 rural, PKR300 or US$1.89 urban) is low for rural projects based on AKAH's recommendation of PKR200–300 (US$1.26–1.89), but is consistent with the median monthly tariff levels reported by WSCs of PKR20.8 or US$0.13 (rural) and PKR250 or US$1.58 (urban). The reluctance of WSCs to implement the recommended tariff collection strategy was noted in 1998 where 11 out of 15 WASEP schemes did not follow AKAH recommendations (Hussain et al. 2000). Heterogeneous communities (median PKR200 or US$1.26) pay more than homogeneous communities (median PKR120 or US$0.76) and satisfaction levels are higher among rural (60%) and homogeneous (58%) communities that pay lower tariffs compared to urban (33%) and heterogeneous (30%) communities that pay much higher tariffs.

Table 8.10 O&M survey responses: community contributions, Gilgit-Baltistan (GB), rural, urban

	GB	*Rural*	*Urban*	*HOM*	*HET*
Sample size	*n = 3,132*	*n = 1,123*	*n = 2,009*	*n = 1,426*	*n = 1,706*
O&M13 Frequency of payment for WASEP water					
At start only (%)	56	52	58	8	28
At start and every week/month (%)	35	36	35	25	1
O&M14 Contribution at the start of WASEP					
Nothing (%)	1.7	2.0	1.5	1.8	1.8
Cash, no. of households (%)	85	89	83	90	81
Cash amount, median (PKR)	3,500	3,000	7,000	3,200	6,400
O&M16 Labour contribution at start					
No. of households (%)	49	84	29	58	41
No. of hours, median	8	8	8	8	8
O&M17 Payment for WASEP each month					
Nothing (%)	39	39	39	28	48
Cash, no. of households (%)	49	55	45	66	34
Cash amount, median (PKR)	200	33	300	120	200
O&M24 Cost of WASEP water					
Bad, very bad (%)	12.7	6.8	16.0	7.4	17.1
Good, very good (%)	43	60	33	58	30

Note: HOM, homogeneous; HET, heterogeneous

Responses across the eight districts generally mirror these findings and reflect the differences between rural and urban projects. Gilgit district, with the majority of WASEP urban DWSS, has the highest upfront cash contributions per household for the O&M fund (median PKR11,000), lowest number of households contributing labour (21%), highest monthly tariff (median PKR300), and lowest satisfaction with the cost of WASEP water (31%) (Table 8.11). Households in poorer districts such as Diamer and Nagar do not pay regular tariffs, and only 2% of households in Astore pay regularly. Of the districts where households pay regular tariffs, the median amount is significantly lower than the recommended range, especially for Skardu (PKR20) and Ghizer (PKR30), with the exception of Gilgit (PKR300). Similarly, the financial contribution per household at the start is noticeably lower than the expected PKR3,000 rural connection fee in Astore (PKR1,300), Nagar (PKR1,500) and Skardu (PKR1,000), with most households paying upfront aside from Nagar (55%).

The unwillingness of households to pay can be traced to household responses to the question of payment based on the quantity of water used (O&M25) where 46% of households overall believed that water should be free (Table 8.12). This view is stronger among rural (55%) households, which is consistent with the difference between homogeneous (50%) and heterogeneous (44%) communities. Objections to (or the lack of support for) payment by quantity is partly reflected in 50% of households overall disagreeing with the introduction of water meters (O&M26), with opposition (or lack of support) again higher in rural (63%) and homogeneous (54%) compared with urban (43%) and heterogeneous (47%). Despite this, there appears to be a significant minority support for water meters overall (42%) and in rural (34.1%), homogeneous (40%), and heterogeneous (44%) projects, with the majority of households (46.5%) in favour of water meters in urban projects.

Objections to payment by quantity (O&M25) are especially prevalent in districts where none or very few households pay regular tariffs (Astore 62%; Diamer 84%; Nagar 84%) and where tariff levels are very low (Skardu 72%) (Table 8.13). The majority of households in these districts also do not agree with the introduction of water meters (O&M26), with households in Diamer (100%), Nagar (80%), and Shigar (70%) especially opposed. Hunza and Gilgit are the only districts where a (small) majority of households support the introduction of water meters.

Households from Jutial projects have lower objections to both payment by quantity (O&M25) (with an average of 36% of households believing that water should be free) compared to homogeneous (50%) and heterogeneous (44%) projects, and to the introduction of water meters (34%) compared with urban (43%) and heterogeneous (47%) (Tables 8.14 and 8.12). Support for payment for water by quantity is highest in Aminabad (68%) followed by Wahdat (28%) and Zulfiqarabad (26%), and lowest in Noorabad (2.5%) and Yasin Colony (2.8%). However, while this is reflected in support for water meters (O&M26) in Aminabad (81% in support), Wahdat (49%), and Zulfiqarabad (70%), there is

Table 8.11 O&M survey responses: community contributions, by district

	Astore	*Diamer*	*Ghizer*	*Gilgit*	*Hunza*	*Nagar*	*Shigar*	*Skardu*
Sample size	*n* = 60	*n* = 32	*n* = 614	*n* = 1,703	*n* = 369	*n* = 58	*n* = 162	*n* = 134
O&M13 Frequency of payment for WASEP water								
At start only (%)	97	100	56	59	18	64	77	67
At start and every week/month (%)	2	0	32	34	76	0	11	23
O&M14 Contribution at the start of WASEP								
Nothing (%)	0.0	0.0	0.5	1.7	0.3	31.0	0.6	0.7
Cash, no. of households (%)	93	100	94	81	89	55	90	83
Cash amount per household, median (PKR)	1,300	3,000	3,000	11,000	4,000	1,500	3,000	1,000
O&M16 Labour contribution at start								
No. of households (%)	88	100	82	21	81	76	88	75
No. of hours (median)	8	8	8	7	8	8	8	8
O&M17 Payment for WASEP each month								
Nothing (%)	73	100	37	38	5	100	70	63
Cash, no. of households (%)	7	0	59	45	89	0	26	23
Cash amount, median (PKR)	100	0	30	300	100	0	120	20
O&M24 Cost of WASEP water								
Bad, very bad (%)	11.7	0.0	9.8	15.4	6.8	5.2	11.7	14.9
Good, very good (%)	65	100	65	31	48	57	39	47

Note: Percentages may not add up to 100% as neutral and non-responses are excluded.

Table 8.12 O&M survey responses: attitudes to tariffs, Gilgit-Baltistan (GB), rural, urban

	GB	*Rural*	*Urban*	*HOM*	*HET*
Sample size	*n* = 3,132	*n* = 1,125	*n* = 2,007	*n* = 1,426	*n* = 1,706
O&M25 Payment by quantity of water used					
No, water should be free (%)	46	55	42	50	44
No, households should all pay the same fee or according to household wealth (%)	34.5	34.5	34.3	36.3	32.8
Yes, households should pay the same price per litre (%)	8.3	4.4	10.5	5.8	10.4
Yes, households who use more water should pay more per litre (%)	5.8	2.4	7.8	3.9	7.5
O&M26 Introduction of water meters					
Yes (%)	28.0	18.7	33.2	25.3	30.2
Yes, if functioning and affordable (%)	14.0	15.4	13.3	14.2	13.8
No (%)	50	63	43	54	47
O&M32 WASEP worth the money and labour					
Agree, strongly agree (%)	57	71	50	68	49
Disagree, strongly disagree (%)	13.6	11.9	14.5	10.5	16.2

Note: HOM, homogeneous; HET, heterogeneous

Table 8.13 O&M survey responses: attitudes to tariffs, by district

	Astore	*Diamer*	*Ghizer*	*Gilgit*	*Hunza*	*Nagar*	*Shigar*	*Skardu*
Sample size	*n* = 60	*n* = 32	*n* = 614	*n* = 1,703	*n* = 369	*n* = 58	*n* = 162	*n* = 134
O&M25 Payment by quantity of water used								
No, water should be free (%)	62	84	55	41	26	84	71	72
No, households should all pay the same fee or according to household wealth (%)	16.7	15.6	39.4	39.1	48.8	10.3	23.5	14.9
Yes, households should pay the same price per litre (%)	10.0	0.0	2.8	10.5	13.6	0.0	1.2	5.2
Yes, households who use more water should pay more per litre (%)	6.7	0.0	0.8	8.2	7.6	3.4	1.9	0.7
O&M26 Introduction of water meters								
Yes (%)	26.7	0.0	16.8	34.2	34.7	10.3	13.0	14.2
Yes, if functioning and affordable (%)	11.7	0.0	18.6	12.4	18.2	5.2	15.4	15.7
No (%)	57	100	62	42	45	84	70	66
O&M32 WASEP worth the money and labour								
Agree, strongly agree (%)	75	81	80	46	66	57	59	55
Disagree, strongly disagree (%)	6.7	0.0	7.8	15.0	13.6	15.5	21.6	17.9

Table 8.14 O&M survey responses: attitudes to tariffs, Jutial projects

	Aminabad	*Astore Colony*	*Diamer Colony*	*Noor Colony*	*Noorabad Extension*	*Wahdat Colony*	*Yasin Colony*	*Zulfiqarabad*
Sample size:	*n* = 77	*n* = 68	*n* = 70	*n* = 80	*n* = 79	*n* = 82	*n* = 72	*n* = 82
O&M25 Payment by quantity of water used								
No, water should be free (%)	14.3	39.7	50.0	30.0	55.7	20.7	41.7	35.4
No, households should all pay the same fee or according to household wealth (%)	6.5	41.2	12.9	42.5	39.2	35.4	47.2	37.8
Yes, households should pay the same price per litre (%)	55.8	16.2	11.4	17.5	0.0	9.8	1.4	19.5
Yes, households who use more water should pay more per litre (%)	11.7	0.0	21.4	10.0	2.5	18.3	1.4	6.1
O&M26 Introduction of water meters								
Yes (%)	51.9	36.8	44.3	41.3	44.3	30.5	30.6	63.4
Yes, if functioning and affordable (%)	28.6	7.4	0.0	7.5	5.1	18.3	11.1	6.1
No (%)	5.2	51.5	22.9	42.5	50.6	20.7	50.0	29.3
O&M32 WASEP worth the money and labour								
Agree, strongly agree (%)	48.1	67.6	21.4	48.8	38.0	22.0	23.6	51.2
Disagree, strongly disagree (%)	1.3	1.5	1.4	3.8	0.0	3.7	4.2	1.2

Table 8.15 O&M survey responses: attitudes to tariffs, Danyore projects

	Amphary Patti (Mbad)	*Chikas Kote*	*Hunza Patti (Mbad)*	*Hussain pura etc.*	*Princeabad Bala etc.*	*Shangote Patti (Mbad)*	*Sharote etc.*	*Sultanabad 1&2*	*Syedabad*
Sample size	*n* = 106	*n* = 89	*n* = 92	*n* = 102	*n* = 91	*n* = 121	*n* = 88	*n* = 119	*n* = 79
O&M25 Payment by quantity of water used									
No, water should be free (%)	39.6	34.8	38.0	32.4	83.5	40.5	45.5	47.9	36.7
No, households should all pay the same fee or according to household wealth (%)	28.3	49.4	37.0	30.4	9.9	36.4	48.9	20.2	40.5
Yes, households should pay the same price per litre (%)	0.0	9.0	0.0	22.5	0.0	17.4	0.0	11.8	8.9
Yes, households who use more water should pay more per litre (%)	10.4	4.5	18.5	5.9	5.5	2.5	2.3	16.8	11.4
Don't know (%)	14.2	2.2	4.3	5.9	0.0	2.5	3.4	1.7	0.0
O&M26 Introduction of water meters									
Yes (%)	17.0	30.3	31.5	27.5	26.4	37.2	31.8	28.6	21.5
Yes, if functioning and affordable (%)	13.2	6.7	12.0	15.7	4.4	12.4	25.0	26.1	21.5
No (%)	54.7	56.2	37.0	48.0	51.6	32.2	42.0	42.0	50.6
Don't know (%)	7.5	6.7	19.6	5.9	17.6	17.4	0.0	1.7	0.0
O&M32 WASEP worth the money and labour									
Agree, strongly agree (%)	47.2	29.2	31.5	17.6	35.2	27.3	33.0	12.6	60.8
Disagree, strongly disagree (%)	4.7	18.0	29.3	6.9	19.8	24.8	15.9	25.2	10.1
Can't decide, don't know (%)	36.8	39.3	39.1	20.6	45.1	38.8	50.0	59.7	27.8

also majority support (Diamer Colony, Noor Colony) or large minority support (Astore Colony, Noorabad Extension, Yasin Colony). This may be in part due to respondents misunderstanding the question on payment by quantity but also suggests that urban households may be more receptive to the introduction of payment by meter.

Compared to Jutial, an average 44% of households in (peri-urban) Danyore believe that water should be free (O&M25), with this figure above 30% across all Danyore projects and highest by far in Princeabad Bala etc. (84%) followed by Sultanabad 1&2 (48%) and Sharote etc. (46%) (Table 8.15). Objection to water meters (O&M26) is also higher on average at 46% compared to 34% in Jutial.

8.6 Conclusion

WASEP has successfully mobilized communities to contribute to and sustain the operations of DWSS. This is reflected in significantly higher community shares of total project costs especially among rural households, costs per person that compare favourably with other developing and lower-income countries, and sustained operations. However, the long-term financial performance of the sampled projects highlights the difficulties of achieving financial and operational sustainability due to the inherent features of the CBWM model. The first is the requirement for regular tariff collection/payment to cover the expenses of O&M that assumes that (poor) households are able and/or willing to pay. The evidence shows that rural households are the least likely to pay regular tariffs and most opposed to the payment for water, despite higher levels of participation and community contributions. Higher community contributions are likely facilitated by in-kind (and not cash) contributions and the lower costs associated with simpler gravity-fed systems. As a result, the value of rural contributions is lower than urban contributions despite accounting for a higher share of total project costs.

The second feature of the CBWM model is the tension between affordable and hence low tariffs and cost recovery; average monthly rural tariff levels of PKR23.44 (US$0.57 based on exchange rates for 2019–2021) in particular are well below the Pakistan average of PKR193 (US$4.69) and AKAH's recommended range of PKR100–500 (US$2.43–12.16) a month. The average urban tariff is significantly higher at PKR195.93 (US$4.77). At the same time, over 30% of rural WSCs do not collect tariffs regularly despite these being set very low, which also means that most, especially rural, WSCs do not employ a full-time WSO for regular O&M, and cannot cover OpEx, with only 25% of functioning rural WSCs estimated to have a monthly operating surplus. Bank statements also indicate that rural O&M funds are potentially being run down in the absence of sufficient and regular tariff collection. The inability of rural households to pay even very low tariffs is reflected in the low support for payment by usage, including the introduction of water meters necessary for long-term sustainability.

Urban WASEP projects have been financially more successful and arguably better managed, most likely due to socioeconomic differences. Urban households are more willing to pay and are also more supportive of the introduction of water meters despite lower satisfaction because of the higher costs. However, there are differences between urban communities: households in Jutial, for example, are more supportive than in peri-urban Danyore despite having to pay more due to the higher fixed and running costs of lifting water from Gilgit River. This suggests that long-term funding aside, some urban projects may be more financially sustainable and scalable, provided there is long-term support for major repairs, asset rehabilitation, and system expansion.

CHAPTER 9
WASEP water infrastructure and water quality

Attaullah Shah, Manzoor Ali, Jeff Tan, and Mushtaque Ahmed

9.1 Introduction

The rationale of community-based water management (CBWM) is that community participation and management is necessary for sustainable water services. Central to this is the principle of community control including 'the ability to make strategic decisions about how a system is designed, implemented, and managed' (Schouten and Moriarty 2003: 165). Successful community management is then about the ability of communities to maintain operations necessary for the sustainability of water services. This emphasis on the 'software' (management) component of community management presupposes that the 'hardware' (water infrastructure) component is already in place and, more crucially, has been properly designed and constructed. However, engineering is the foundation of sustainable water services because well-designed and constructed water infrastructure is necessary not just for clean drinking water but also to ensure long-term functionality and hence sustainable operations.

The strong emphasis on engineering, infrastructure, and water quality distinguishes the Water and Sanitation Extension Programme (WASEP) from the CBWM model and the focus in the wider literature on 'software'. Not only is WASEP initiated by the implementing NGO, but all technical and engineering aspects of the project are also led by the Aga Khan Agency for Habitat (AKAH), with AKAH engineers making all technical decisions around system design contrary to the CBWM principle of community control of all decision-making. This emphasis on engineering has arguably laid the foundations for the successful delivery of drinking water supply schemes (DWSS) in Gilgit-Baltistan in terms of functionality and sustained operations. However, sustained operations also depend on successful operations and maintenance (O&M) and, as such, engineering cannot be separated from community capacities to manage water projects.

This chapter examines the key features of WASEP's engineering-led approach to assess the importance of engineering in the design and construction of water infrastructure and delivery of clean drinking water. Section 9.2 provides a brief overview of DWSS in Gilgit-Baltistan and Gilgit City to help situate and identify the key WASEP engineering features in Section 9.3. Section 9.4 examines how infrastructure quality impacts water quality, looking at infrastructure and water quality scores from an engineering audit of WASEP and non-WASEP (control) sites, and water quality tests. Section 9.5 discusses the implications of these results and concludes.

9.2 Drinking water supply systems in Gilgit-Baltistan

Access to water is not an issue in Gilgit-Baltistan where glaciers provide 50.5 billion cubic metres of water annually to the river Indus (Hussain et al. 2022: 920). Furthermore, 79% of the population (95% urban, 76% in rural) has access to improved water sources (e.g. piped water on premises or public taps) although this varies across districts from 96.5% (Hunza) to 47.3% (Diamer), with significant water shortages in urban areas during the winter (GoP 2019). Surface water (springs, rivers, and streams) is the main source of drinking water and piped water to yards (37%) and houses (26%), with groundwater (bore holes) accounting for only 1% (GoP 2019).

The two main types of piped water systems in Gilgit-Baltistan are gravity-fed and mechanized. The majority of piped systems, especially in rural areas, are gravity-fed where raw water is transferred from springs or streams and rivers through artificial channels to a storage reservoir and distribution network. While urban settlements established on slopes feature piped systems operated by gravity, urban water systems tend to be mechanized, especially if communities do not have water rights to closer water sources, and necessitates pumping water from rivers/streams to storage tanks. Rural DWSS are maintained by the Local Government and Rural Development Department (LG&RDD) while urban DWSS are constructed and operated by the Public Health Engineering Department (PHED), Gilgit Development Authority (GDA), and Public Works Department (PWD) (GB-EPA 2019).

In contrast to the availability of water, water quality has always been poor and appears to be deteriorating in Gilgit-Baltistan; 69% of the population are without access to safely managed drinking water (GoP 2019) and 78% of water sources are unsafe 'mainly due to the prevalence of microbial contamination' (Ahsan et al. 2021: v). Maintaining drinking water quality 'remains an issue in almost all urban areas' due to the lack of proper design and treatment facilities, insufficient treatment capacity to meet population growth, understaffing, underfunding, a shortage of power to run pumps, the lack of spare parts, and the absence of a responsible government agency (GB-EPA 2012: 12). None of the public water systems in urban centres are treated, with water pumped directly from source to storage reservoirs without primary or secondary treatment (GB-EPA 2013).

The main urban centre in Gilgit-Baltistan is Gilgit City, where Gilgit River and Kargah and Jutial Nallahs (streams) are the main sources for drinking water supply networks. Kargah Nallah supplies 70% of water for Gilgit town and is distributed from Barmas Complex. Jutial Nallah is the main source of water for Jutial and Khomar. Source water from Kargah and Jutial Nallah is brought to Jutial and Barmas water complexes through an unfenced constructed water channel. Urban populations settled in Konodas and Zulfiqarabad rely mostly on water from Gilgit River for drinking and irrigation. River water is lifted through pumps to water storage reservoirs situated at both ends of the river and is distributed to households through gravity (GB-EPA 2012).

In addition to the PHED water supply networks, two community-owned water pumps are also operational in Jutial. These water pumps fulfil the requirements of new settlements in Jutial and Zulfiqarabad. Despite the PHED and community-owned water supply systems, Gilgit City faces a severe shortage of water in the spring and summer (due to the mis-use of drinking water for irrigation), with households having to resort to paying for water tankers. The lack of sanitation systems and environmental legislation means that all types of surface runoff and waste effluent from households and commercial areas directly enter nearby water collecting bodies and subsequently Gilgit River (GB-EPA 2012).

9.3 WASEP engineering features

The focus of WASEP on engineering can be illustrated in its standard operating procedures (SOPs) and water infrastructure. The aim of the SOPs is to ensure the uninterrupted supply of safe drinking water through user-friendly designs, the use and proper installation of quality materials, and 'a robust monitoring mechanism' (AKPBS-P 2012: 58). The engineering process begins with (1) the selection of potential source(s), informed by input from the local community, and based on yield, accessibility, distance, the type of source and type of scheme, and apparent quality of water. The source is selected if water quality meets World Health Organization (WHO) standards for safe drinking water and the yield is sufficient. This is followed by (2) a topographic survey of water yield; GPS readings of source(s), tank, bottom drains, air release valves, and prominent points (schools, places of worship, offices); vertical angles for feeder/supply pipes and distribution network; length of pipes and *nallah* (stream/watercourse) crossings; soil conditions (every 100 m); population and projected growth rate; and availability of local materials and transportation costs (AKPBS-P 2012) (see Figure 9.1).

The field survey data is then (3) analysed to confirm vertical angles, length, and elevation, and (4) the water distribution network modelled, based on calculations of average and peak demand with simulations to adjust pipe diameters for optimum water pressure. The SOPs also specify detailed procedures for the location and design of civil structures (intake chamber, storage tank, sedimentation tank, upflow roughing filters, slow sand filters, valve box, distribution chamber, sump, *nallah* crossing, pump

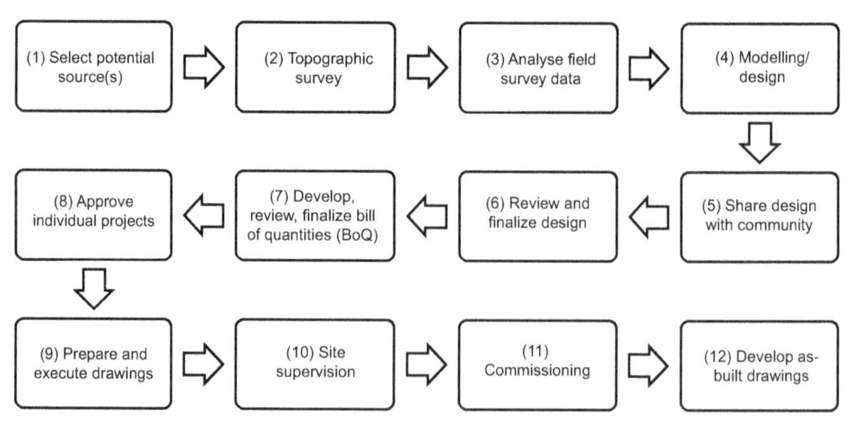

Figure 9.1 WASEP engineering SOPs, flow chart
Source: AKPBS-P (2012).

room). AKAH places a high priority on the quality of water at the source, with a water intake chamber (for spring water sources) protected from human and animal contamination. A storage tank is needed if the yield is not sufficient to meet the peak hour demand or if the distance to the source is greater than 2 km. The storage tank and distribution chamber are located following a hazard risk assessment and protected from surface runoff (AKPBS-P 2012).

The water system design is shared with the community through (5) a general body meeting where feedback is sought, with 'genuine' or valid changes incorporated, and the design signed off by the water and sanitation committee. The design is then (6) reviewed and finalized by AKAH senior engineers and managers with peer reviews conducted in certain 'distinct' cases. Project costing is conducted through (7) a bill of quantities (BoQ) that records the quantity and cost of materials and labour. This covers the length of rock hanging and *nallah* crossing; actual cost of transportation of material, both local (sand, gravel, and stone) and non-local (cement, pipes, mild steel bars, etc.); an O&M fund and contingency fund (of at least 1% of total project cost); size of tank and pump based on average daily demand; and blasting costs, all based on existing market prices (AKPBS-P 2012).

Individual projects are then (8) approved following the endorsement of the BoQ by AKAH management, and (9) drawings are prepared in AutoCAD of the water distribution network, civil structures, sewage collection network and appurtenances, process flow diagram for pumping scheme, electrification, pump house, and sump. The construction of the system involves (10) site supervision with monitoring by a trained site supervisor and mandatory site visits and log reports by the AKAH site engineer, senior field engineer, and technical manager. The commissioning stage (11) involves regulating the flow through control valves to ensure equal distribution and required pressure to each household, and addressing any problems such as leaks (AKPBS-P 2012). The final step is (12) developing as-built drawings on AutoCAD incorporating any changes to the original design.

A number of notable water infrastructure features emerge from the WASEP SOPs. The locations for civil structures (e.g. intake chamber, reinforced concrete storage tank, sedimentation tank, upflow roughing filters, slow sand filters, valve box, distribution chamber, sump) are based on a hazard risk assessment and protected as much as possible from natural hazards through careful site selection, with remedial protective measures taken if required. Protection from surface runoff is provided by galvanized iron covers with benching, with watertight structures for the intake chamber, sedimentation tank, upflow roughing filters, and distribution chamber. Distribution pipes are laid 4 ft (1.2 m) underground and below the frost line to prevent pipes from freezing and bursting. Similarly, riser pipes connecting the water mains to household tap stands are laid 3 ft (0.9 m) underground and the cylindrical concrete casing filled with sawdust for additional thermal insulation.

The pipes themselves are made from HDPE with compression fittings, and straw-insulated steles are used at the household level. Where no clean spring water is available, AKAH either incorporates percolating systems such as a sump and well for the supply of river filtrate, or gravity flow percolation systems for the supply of usually not clean stream or river water (AKPBS-P n.d.). AKAH selects and procures external material (e.g. HDPE pipes, blasting materials, fittings, cement, steel) based on donor funding and the community is responsible for providing or paying for labour and local materials (land, stones, sand, etc.).

As described in the SOPs, water quality is tested during selection of the source and again at the source, system, and tap upon project completion. This is to ensure that water quality at source, system, and tap meets WHO standards for safe drinking water (category A) before handover to the community. In other words, the water quality in all WASEP projects is technically safe for drinking at the point of project handover.

9.4 WASEP engineering and water quality: evidence

The engineering evidence is drawn from engineering audits and water quality tests from a sub-sample of 12 rural and 14 urban projects. The engineering sub-sample was generated from the larger research sample of 25 rural and 25 urban projects and based on variations of engineering features (e.g. mechanical/gravity-fed water system, pipe length, number of household connections) and then quota sampled by district. Six rural and six urban (non-WASEP) control sites were then selected to match these engineering features in the same districts (see Tables 9.1 and 9.2).

The expected lifespan of rural and urban water infrastructure is 15 and 20 years respectively. This means that four (30%) rural WASEP projects in the engineering sub-sample were operating beyond their natural lifespans, with

Table 9.1 Sample overview, rural WASEP and control

Site (rural)	*District*	*District group*	*Source*	*Water system (G/M)*	*Water samples (no.)*	*Project completion (year)*	*Project age (years)*
WASEP:							
Broshal	Nagar	Hunza-Nagar	Spring	G	3	1998	24
Chandupa*	Shigar	Baltistan	Spring	G	N/A	2000	10
Daeen Chota	Ghizer	Ghizer	Spring	G	4	2006	16
Duskhore Hashupi	Shigar	Baltistan	Spring	G	3	2008	14
Dushkin	Astore	Astore	Spring	G	3	2006	16
Hatoon Paeen	Ghizer	Ghizer	Spring	G	3	2011	11
Kirmin	Hunza	Hunza-Nagar	Spring	G	5	2018	4
Nasir Abad	Ghizer	Ghizer	Spring	G	4	2011	11
Rahimabad (M'dass)	Gilgit	Gilgit	Spring	G	4	2012	10
Shamaran Paeen	Ghizer	Ghizer	Spring	G	3	2008	14
Shilati	Diamer	Diamer	Spring	G	3	2013	9
Singul Shyodass	Ghizer	Ghizer	Spring	G	4	2004	18
Control:							
Doyan	Astore	Astore	Spring	G	4	2014	8
Thorgu Paeen	Skardu	Baltistan	Stream	G	0	2017	5
Samigal	Darel	Diamer	Stream	G	4	2014	8
Damas	Punial	Ghizer	Spring/ stream	G	3	1986	36
Jutal	Gilgit	Gilgit	Stream	G	3	2014	8
Hussainabad	Hunza	Hunza-Nagar	Stream	G	3	2014	8

Notes: Water system covers G (gravity) and M (mechanical). Project age is taken at 2022. *Chandupa was only operational until 2010. Water samples could not be collected from Chandupa as the project was not functional.

a further two close to this at 14 years (Table 9.1). In contrast urban WASEP projects are much newer, having only being recently introduced as a result of urbanization and requests by the Government of Gilgit-Baltistan, with an average age of 6.9 years for the engineering sub-sample (Table 9.2).

9.4.1 Water infrastructure

Water infrastructure quality is based on scores awarded by the research team during engineering audits of WASEP and control sites. The engineering audit involved the physical inspection of source water intake, water supply route,

Table 9.2 Sample overview, urban WASEP and control

Site (urban)	*Union*	*District group*	*Source*	*Water system (G/M)*	*Water samples (no.)*	*Project completion (year)*	*Project age (years)*
WASEP:							
Aliabad Centre	Hunza	Hunza-Nagar	Spring	G	3	2014	8
Aminabad	Jutial	Gilgit	Sump	M	3	2013	9
Astore Colony	Jutial	Gilgit	Sump	M	3	2018	4
Chokoporo Gahkuch Bala	Municipal area	Ghizer	Spring	G	3	2015	7
Diamer Colony	Jutial	Gilgit	Sump	M	3	2018	4
Domial Buridur Gahkuch Bala	Municipal area	Ghizer	Spring	G	3	2015	7
Khanabad	Municipal area	Ghizer	Spring	G	3	2015	7
Noor Colony Urban Project	Gilgit	Gilgit	Sump	M	3	2014	8
Noorabad Extension	Jutial	Gilgit	Sump	M	3	2014	8
Rahimabad Aliabad	Aliabad	Hunza-Nagar	Stream	G	3	2012	10
Soni Kot	Gilgit	Gilgit	Sump	M	3	2011	11
Wahdat Colony	Jutial	Gilgit	Sump	M	3	2018	4
Yasin Colony	Jutial	Gilgit	Sump	M	3	2018	4
Zulfiqarabad	Jutial	Gilgit	Sump	M	3	2016	6
Control:							
Gahkuch Khari	Gahkuch	Ghizer	Sump	M	2	1987	35
Jutial Kot Mohallah	Gilgit	Gilgit	Stream	G	3	1984	38
Jagir Baseen	Gilgit	Gilgit	Stream	G	3	1988	34
Konodas	Gilgit	Gilgit	River	G	2	1984	38
Sikandarabad	Nagar	Hunza-Nagar	Stream	G	3	1992	30
Sakwar	Gilgit	Gilgit	Stream	G	3	1984	38

Notes: Water system covers G (gravity) and M (mechanical). Project age is taken at 2022.

and water distribution network. Each part of the water infrastructure network was assessed in terms of risk of contamination and damage from natural hazards, and scored on a Likert scale of 1 (very poor), 2 (poor), 3 (satisfactory), 4 (good), and 5 (very good). Higher scores were awarded for water sources, water supply routes, and distribution networks that were well designed, secure, and protected from contamination from human or livestock activity, and from natural hazards. Similarly, given the problem of freezing pipes in winter, water supply routes and water distribution networks were scored higher where pipes were buried below the frost line, which had the added benefit of reducing illegal water connections and the accompanying risk of contamination. Finally, the percentage of pipes with weak joints and leaks was factored into water supply route scores.

Rural WASEP schemes recorded higher infrastructure scores at source intake (average 4.3 out of 5), route (4.2), and distribution network (4.0). In comparison the average scores for rural control sites were 2.7 (source intake), 3.3 (route), and 1.3 (distribution network) (Table 9.3). The differences in average infrastructure scores were even higher for (newer) urban WASEP projects: source

Table 9.3 Water infrastructure scores, rural WASEP and control

Site (rural)	*Infrastructure score: source intake*	*Infrastructure score: route*	*Infrastructure score: distribution network*	*Infrastructure score: average*
WASEP:				
Broshal	3	4	4	3.7
Chandupa	3	2	3	2.7
Daeen Chota	4	4	4	4.0
Duskhore Hashupi	5	5	5	5.0
Dushkin	5	5	4	4.7
Hatoon Paeen	4	4	4	4.0
Kirmin	3	1	4	2.7
Nazir Abad	5	5	5	5.0
Rahimabad (Matumdass)	5	5	2	4.0
Shamaran Paeen	4	5	4	4.5
Shilati	5	5	5	5.0
Singul Shyodass	5	5	4	4.7
Average:	*4.3*	*4.2*	*4.0*	*4.2*
Control:				
Doyan	2	3	1	2
Thorgu Paeen	2	4	1	2.3
Samigal	3	3	1	2.3
Damas	4	3	1	2.7
Jutal	2	3	2	2.3
Hussainabad	3	4	2	3
Average:	*2.7*	*3.3*	*1.3*	*2.4*

Notes: Water infrastructure scores are out of 5, with 1 (very poor), 2 (poor), 3 (satisfactory), 4 (good), and 5 (very good).

Table 9.4 Water infrastructure scores, urban WASEP and control

Site (urban)	*Infrastructure score: source intake*	*Infrastructure score: route*	*Infrastructure score: distribution network*	*Infrastructure score: average*
WASEP:				
Aliabad Centre	2	5	5	4.0
Aminabad	5	5	5	5.0
Astore Colony	5	5	5	5.0
Chokoporo Gahkuch Bala	5	5	4	4.7
Diamer Colony	5	5	5	5.0
Domial Buridur Gahkuch Bala	5	5	4	4.7
Khanabad Gahkuch Bala	5	5	4	4.7
Noor Colony	5	5	5	5.0
Noorabad Extension	5	5	5	5.0
Rahimabad Aliabad	2	5	5	4.0
Soni Kot	5	5	5	5.0
Wahdat Colony	5	5	5	5.0
Yasin Colony	5	5	5	5.0
Zulfiqarabad	5	5	5	5.0
Average:	*4.6*	*5.0*	*4.8*	*4.8*
Control:				
Gahkuch Khari bazaar area	N/A*	N/A*	N/A*	N/A*
Jutial Kot Mohallah	1	3	1	1.7
Jagir Baseen	4	3	1	2.7
Konodas	3	3	0	2.0
Sikandarabad	0	3	1	1.3
Sakwar	1	4	1	2.0
Average:	*1.8*	*3.2*	*0.8*	*1.9*

Notes: Water infrastructure scores are out of 5, with 1 (very poor), 2 (poor), 3 (satisfactory), 4 (good), and 5 (very good). N/A* (no water infrastructure).

intake (WASEP 4.6, control 1.8), route (WASEP 5.0, control 3.2), and distribution network (WASEP 4.8, control 0.8) (Table 9.4). The engineering audit thus confirms the higher quality of WASEP water infrastructure.

9.4.2 Water quality

A total of 56 water samples were collected from 18 rural sites – 39 samples from 12 WASEP projects and 17 samples from 6 control sites (Table 9.1). Rural water samples were collected between August and October 2020 and urban water samples were collected in March and April 2021, at source, system (storage or distribution), and tap. Water samples were not collected in the winter as the cold or freezing water masks microbiological contamination. Water samples were tested on site for microbiological and physicochemical parameters and transported to AKAH's lab for chemical tests. All rural sites were gravity-based, relying on spring or stream (*nallah*) water. All WASEP water sources were from springs while only one control site relied solely on a spring water source. Both rural WASEP and control sites were around the same average age (14.1 years WASEP, 12.2 years control). However, once the outlier Damas (36 years) is removed from control sites, the average age of control sites is around half the age of WASEP at 7.4 years.

A total of 58 water samples were collected from 20 urban sites – 42 from 14 WASEP projects and 16 from 6 control sites (Table 9.2). In contrast to rural projects, 64% of urban WASEP projects were mechanical systems, relying on water pumped from rivers. Urban WASEP projects are also significantly younger than rural WASEP and urban control projects, having only been recently introduced and with an average age of 6.9 years compared to 35.5 years for control sites.

Although rural WASEP projects are older, the overall water quality based on *E. coli* colonies was in the range of safe and low risk, with 100% WASEP water samples at the source and 92% at system and tap coming under category A (safe drinking water). In comparison, no water samples from control sites came under category A, with 100% of collected water samples at source, 80% at system, and 40% at tap falling under category B (low risk) and the rest in category C (medium risk) (Table 9.5). Water samples could not be collected from Chandupa as the project was not functional.

In the case of Broshal, the only WASEP site with scores below category A, the deterioration of water quality was due to a combination of age and increasing demand. At 24 years, Broshal had already significantly exceeded the 15-year lifespan for rural water infrastructure. Wear and tear along with land movement had resulted in damage to joints, leaks, and the subsequent contamination of water in the system. At the same time, population growth along with an increase in tourism and the construction of houses, guest houses, and hotels meant that demand had risen beyond the system's capacity. To meet this increased demand, the community added water from an open stream into the main water reservoir and in doing so introduced *E. coli* into the water system.

As with the rural sample, microbial water quality was significantly better in urban WASEP projects with 86% (at source), 75% (system), and 86% (tap) of water samples falling under category A. This compares with no water samples testing in category A for control sites and 33% (source), 0% (system), and 17% tap falling under category B, with the remainder in category C (Table 9.6).

T-tests to confirm these findings could only be run for rural (system) and urban (source, tap) because of zero variance for rural (source, tap) and urban (system). The differences between *E. coli* water quality scores in rural WASEP

Table 9.5 Water quality (microbiological), rural WASEP and control

Site (rural)	*Microbial: source*	*Microbial: system*	*Microbial: tap*
WASEP:			
Broshal	A	B	C
Chandupa	N/A*	N/A*	N/A*
Daeen Chota	A	A	A
Duskhore Hashupi	A	A	A
Dushkin	A	A	A
Hatoon Paeen	A	A	A
Kirmin	A	A	A
Nazir Abad	A	A	A
Rahimabad (Matumdass)	A	A	A
Shamaran Paeen	A	A	A
Shilati	A	A	A
Singul Shyodass	A	A	A
Control:			
Doyan	B	C	C
Thorgu Paeen	N/A**	N/A**	N/A**
Samigal	B	B	B
Damas	B	B	B
Jutal	B	B	C
Hussainabad	B	B	C

Notes: Category A (safe drinking water), category B (low risk), category C (medium risk). N/A* (water samples could not be collected from Chandupa as the project was not functional); N/A** (site too far to transport water sample to the lab in time).

and control sites were statistically significant based on t-tests for rural (system) and urban (source, tap) (Table 9.7).

The physicochemical values for rural WASEP water were on average higher than at control sites, aside from turbidity: pH (7.5 WASEP, 7.3 control), total dissolved solids (TDS) (165 WASEP, 119 control), conductivity (343 WASEP, 249 control), salinity (155 WASEP, 120 control), total hardness (136 WASEP, 123 control), turbidity (0.7 WASEP, 1.9 control) (Table 9.8). These average values were generally within WHO recommendations although three WASEP projects (Dushkin, Kirmin, Rahimabad) and one control site (Jutal) exceeded WHO guidelines for conductivity, three WASEP projects (Daeen Chota, Dushkin, Kirmin) and two control sites (Jutal, Hussainabad) had very hard water, and one control site (Hussainabad) greatly exceeded turbidity limits (Table 9.8). These higher values are normally attributable to seasonal fluctuations when high temperatures in summer create floods leading to erosions of riverbanks and subsequent high TDS, turbidity, and conductivity. Microbiological and physicochemical contamination in drinking water also depend on season,

Table 9.6 Water quality (microbiological), urban WASEP and control

Site (urban)	*Microbial: source*	*Microbial: system*	*Microbial: tap*
WASEP:			
Aliabad Centre	B	B	B
Aminabad	A	A	A
Astore Colony	A	N/A*	A
Chokoporo Gahkuch Bala	A	A	A
Diamer Colony	A	A	A
Domial Buridur Gahkuch Bala	A	A	A
Khanabad	A	A	A
Noor Colony	A	N/A*	A
Noorabad Extension	A	N/A*	A
Rahimabad Aliabad	B	B	B
Soni Kot	A	N/A*	A
Wahdat Colony	A	N/A*	A
Yasin Colony	A	A	A
Zulfiqarabad	A	N/A*	A
Control:			
Gahkuch Khari	B	N/A**	B
Jutial Kot	B	C	C
Jagir Baseen	C	C	C
Konodas	C	N/A*	C
Sikandarabad	C	C	C
Sakwar	C	C	C

Notes: Category A (safe drinking water), category B (low risk), category C (medium risk). N/A* not applicable as water is supplied to consumers directly from sump through pumps and no water tank is required; N/A** water is supplied from a shallow well through pumps with no water tank.

Table 9.7 Water quality (microbiological), t-test results, selected WASEP and control

Water source	*T-value*	*df*	*P-value*	*95% confidence interval*	*Mean*
Rural system	–5.0484	5.7259	0.00267700	–1.6529578 –0.5652241	2.909091 (WASEP) 1.800000 (control)
Urban source	–6.5657	7.2192	0.00027520	–2.0692446 –0.9783745	2.857143 (WASEP) 1.333333 (control)
Urban tap	–8.7651	8.5861	0.00001438	–2.1299940 –1.2509580	2.857143 (WASEP) 1.166667 (control)

Table 9.8 Water quality (physicochemical), rural WASEP and control

Site (rural)	*pH*	*Total dissolved solids*	*Conductivity*	*Salinity*	*Total hardness*	*Turbidity*
WASEP:						
Broshal	7.6	130	268	200	65	0
Chandupa	N/A	N/A	N/A	N/A	N/A	N/A
Daeen Chota	7.8	173	360	200	210	0
Duskhore Hashupi	7.2	82	172	100	75	0
Dushkin	7.8	280	578*	300*	225	0
Hatoon Paeen	7.6	111	233	100	90	2.3
Kirmin	7.6	512*	1042*	200	450	3.26
Nazir Abad	7.4	47	99	50	60	0
Rahimabad (Matumdass)	7.2	311*	667*	300*	65	0
Shamaran Paeen	7.6	59	126	100	60	1.5
Shilati	7.4	61	129	100	60	0
Singul Shyodass	7.6	46	99	50	60	0.26
Control:						
Doyan	6.8	42	91	0	60	0
Thorgu Paeen						
Samigal	6.8	68	142	100	60	0
Damas	7.6	89	187	100	75	0
Jutal	7.8	233	481*	200	210	0
Hussainabad	7.4	165	342	200	210	9.65*
WHO standard	<8.0	<300	<400	200	10–500	<5

Notes: *Exceeds WHO recommendations for drinking water. Water samples could not be collected from Chandupa as the project was not functional.

as these tend to be higher in summer and reduce gradually in autumn (GB-EPA 2019).

Physicochemical properties were more mixed for urban WASEP and control projects, with higher average values for WASEP for pH (7.8 vs. 7.6 for control) and conductivity (324 vs. 275), and lower values for WASEP for total hardness (119 vs. 139) and turbidity (2.9 vs. 3.9). These average values were within WHO recommendations but, as with several rural projects, 39% of individual WASEP urban projects (Noor Colony, Noorabad Extension, Soni Kot, Wahdat Colony, Zulfiqarabad) and 17% of control sites (Konodas) exceeded WHO guidelines for conductivity, and 8% of WASEP (Zulfiqarabad) and 17% of control sites (Gahkuch Khari) recorded very hard water. High turbidity was also recorded at 14% of WASEP projects (Aliabad Centre (15), Rahimabad Aliabad (15)) and 33% of control sites (Sikandarabad (10), Sakwar (8.4)) compared to WHO recommendations of 5 NTU or less (Table 9.9).

Table 9.9 Water quality (physicochemical), urban WASEP and control

Site (urban)	*pH*	*Total dissolved solids*	*Conductivity*	*Salinity*	*Total hardness*	*Turbidity*
WASEP:						
Aliabad Centre	7.8	144	299	0.1	165	15*
Aminabad	7.6	97	203	0.1	125	5
Astore Colony	7.8	164	340	0.2	105	0
Chokoporo Gahkuch Bala	7.8	40	86	0	45	0
Diamer Colony	7.8	164	340	0.2	105	0
Domial Buridur Gahkuch Bala	7.8	40	86	0	45	0
Khanabad	7.8	40	86	0	45	0
Noor Colony	7.6	268	553*	0.3	150	0
Noorabad Extension	7.6	233	482*	0.2	120	0
Rahimabad Aliabad	7.8	144	299	0.1	165	15*
Soni Kot	7.6	201	416*	0.2	125	5
Wahdat Colony	8.2*	238	491*	0.2	175	0
Yasin Colony	7.8	164	340	0.2	105	0
Zulfiqarabad	7.6	251	518*	0.2	190	0
Control:						
Gahkuch Khari	7.6	166	344	0.2	330	5
Jutial Kot	7.5	84	176	0.1	60	0
Jagir Baseen	7.5	41	88	0	40	0
Konodas	7.6	275	567*	0.3	180	0
Sikandarabad	7.6	153	318	0.1	150	10*
Sakwar	7.5	75	157	0.1	75	8.4*
WHO standard	<8.0	<300	<400	200	10-500	<5

Note: *Exceeds WHO recommendations for drinking water.

9.5 Conclusion

WASEP's integrated approach was introduced following the failure of earlier engineering solutions that simply relied on the provision of clean drinking water to reduce waterborne diseases. However, engineering and 'hardware' (water infrastructure) remains a central feature of WASEP's integrated approach in contrast to the emphasis in the literature on the 'software' (management) component of CBWM. Moreover, instead of community control of strategic decisions including system design and implementation, which is a key principle of CBWM, AKAH retains control over all engineering aspects of water infrastructure even if this is in consultation with the community.

Compared to public water systems that lack treatment, WASEP water infrastructure is guided by engineering SOPs to identify and protect clean water sources, and safeguard water quality through proper design and the use of durable materials. This is reflected in the quality of engineering and water quality, where WASEP projects scored higher in the engineering audit and water quality tests compared with control sites. It can also be seen in sustained operations in rural projects where many schemes have functioned close to or beyond the natural lifespan of water infrastructure despite the irregular collection of tariffs discussed in Chapter 8.

The case of Broshal helps illustrate both the strengths and the weakness of the WASEP model. On the one hand, the continued operation of water services is testament to the underlying quality of WASEP's engineering focus. On the other hand, poorer water quality in Broshal due to the addition of contaminated open stream water to meet growing demand highlights the limitations of financing under the CBWM model, where long-term funds are not available or budgeted for system rehabilitation and expansion.

CHAPTER 10

Natural hazards: WASEP engineering solutions and community responses

Karamat Ali, Jeff Tan, and Manzoor Ali

10.1 Introduction

One of the main challenges in sustaining the Water and Sanitation Extension Programme (WASEP) water services in Gilgit-Baltistan is the difficulty most water and sanitation committees (WSCs) face in undertaking major repairs in the context of a harsh mountainous environment prone to multiple natural hazards such as earthquakes, floods, avalanches, landslides, and glacial movement. This chapter looks at the role of engineering design in sustaining the delivery of water services in the context of risks posed by natural hazards in mountainous regions such as Gilgit-Baltistan, and the ability of communities to respond to the often-inevitable disruptions to water services. It examines the standard operating procedures (SOPs) that inform the design, engineering, and construction of WASEP water infrastructure, and how these are reflected in the results of an engineering audit of WASEP and control sites in terms of days lost from disruptions due to natural hazards. As engineering design can only mitigate but not prevent the impact of natural hazards, the response of WSCs and communities becomes critical for the resumption and sustainable operations of water services.

The efficacy of WSC responses, mainly through self-help initiatives including additional community financial contributions, was important in the absence of external support from the state or the Aga Khan Agency for Habitat (AKAH), the implementing NGO. Coping capacity is thus an important feature of communities that are better able to respond to damage from natural hazards and restore water services. This can in turn be informed by district-level vulnerability indicators (literacy, income, and poverty) that frame the education and occupation profiles of WASEP communities based on household surveys; these serve as vulnerability proxies (see Chapter 2 on background and methodology). While the ability of communities to respond to disruptions from natural hazards is also determined by the impact or severity of natural hazards, the capacity of WSCs to respond appears to be only loosely connected to the vulnerability proxies of communities. Instead, external support to undertake repairs to water infrastructure may be a more critical factor for sustaining water services in the context of the threats of natural hazards.

Section 10.2 identifies the main natural hazards in Gilgit-Baltistan, introduces hazard and vulnerability scores to help determine the level of risk faced in the 10 districts, and locates the engineering sub-samples for WASEP and control sites within the natural hazard profile of each district. Section 10.3 describes WASEP SOPs and WASEP engineering features, presents the infrastructure scores for WASEP and control sites, and examines how these are related to the damage from natural hazards in terms of the number of days lost from damaged water infrastructure. Section 10.4 assesses community and WSC responses to the impact of natural hazards on water infrastructure, and the coping capacity of WSCs to repair damaged water infrastructure and to sustain water services based on proxy vulnerability indicators (educational attainment and occupation). The final section concludes with some overall observations on the challenges of sustaining water services in the context of natural hazards in mountainous regions such as Gilgit-Baltistan.

10.2 Natural hazards in Gilgit-Baltistan

The province of Gilgit-Baltistan in northern Pakistan is located in the western part of three mountain ranges – the Hindu Kush, Karakoram, and Himalayas – and borders Afghanistan, China, and Indian-administered Kashmir (Figure 10.1). Nine of the 10 districts in the province are located along the Karakoram range (Ghizer, Gilgit, Hunza, Nagar, Shigar, and Ghanche; plus parts of Kharmang and Diamer), Himalayan range (Astore and Skardu), and Hindu Kush (Diamer), with snow and glacial melt water in summer the main sources of water flowing into springs, streams, and rivers. Its geographic location around three mountain ranges and with hundreds of peaks above 6,000 m and 7,000 m make this region prone to glacier-related and other natural hazards. Steep mountain peaks pose geological hazards such as avalanches, glacial lake outburst floods (GLOFs), and landslides. Surging and retreating glaciers create unstable glacier lakes (Hewitt 2014; Bhambri et al. 2017; Gilany et al. 2020, all cited in Baig et al. 2021), with flooding, GLOFs and mudslides also triggered by atmospheric hazards and new seasonal low depressions associated with global warming (Baig et al. 2020, cited in Baig et al. 2021). Glacier-induced flash floods are a permanent feature in Hunza, Shigar, and Skardu districts, with Astore extremely vulnerable to flash and river floods during spring, summer, and winter (Easterling et al. 2000, cited in Baig et al. 2021). Gilgit-Baltistan also lies on a number of fault lines and the potential adverse impact caused by seismic zones is high, with almost all of the province lying within moderate seismic zones and parts of Ghizer susceptible to major tectonic movements (Baig et al. 2021).

The risks posed by natural hazards in northern Pakistan can be weighted by the sources of hazard: (a) geology (25%, as the main trigger of earthquakes, mudslides, and landslides); (b) soil composition (15%, as a factor in erosion); (c) seismic zone (12%); (d) fault line (12%); (e) earthquake incidences (11%); (f) elevation (10%); (g) land cover (10%); and (h) slope (5%, as a factor in flooding, avalanches, and landslides) (Baig et al. 2021). Based on this, a total hazard score (from 1 to 10) can be derived for the 10 districts in Gilgit-Baltistan

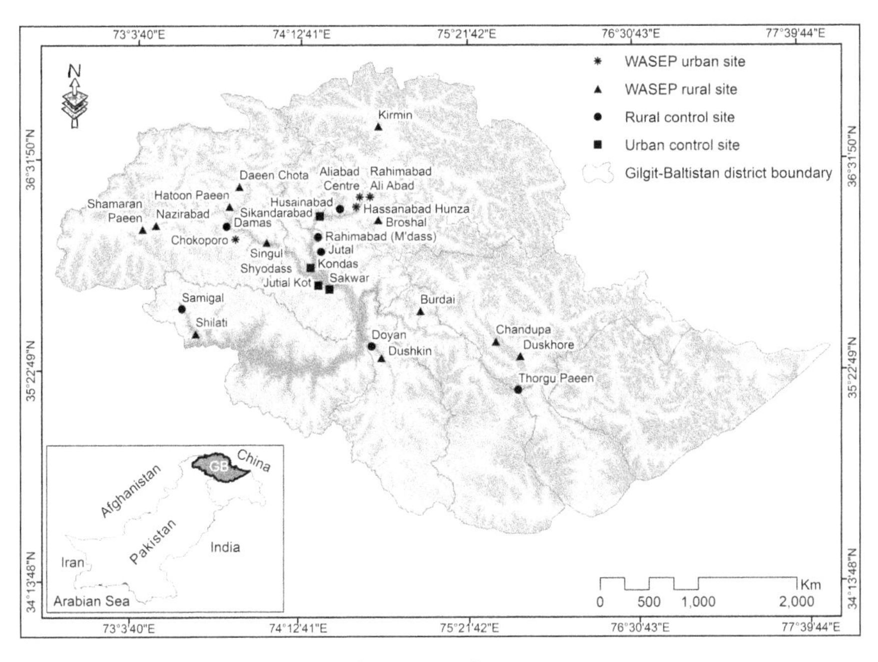

Figure 10.1 Gilgit-Baltistan: Engineering sample sites
Source: map by Karamat Ali and Garee Khan

ranging from 2.75 (low) in Kharmang to 6.85 (high) in Gilgit (Table 10.1). The 12 rural and 16 urban WASEP engineering sub-sample sites along with the six rural and six urban control sites can be located within 8 of the 10 districts in order to provide an overview of the natural hazards facing each site (Table 10.1).

Table 10.1 Gilgit-Baltistan: total risk, district, 2021

District	*Natural hazards*	*Sample sites: control (c), rural (R), urban (U)*	*Total hazard score*
Astore	Flash floods, avalanches, landslides, earthquakes	R: Dushkin R: Doyan (c)	4.78
Diamer	Landslides, earthquakes, floods	R: Shilati R: Samigal (c)	6.48
Ghanche	Landslides, floods, earthquakes	None	7.40
Ghizer	Floods, GLOFs, rockslides, avalanches, land slips, earthquakes	R: Daeen Chota R: Hatoon Paeen R: Nazirabad R: Shamaran Paeen R: Singul Shyodass R: Damas (c) U: Chokoporo Gahkuch Bala U: Domial Buridur Gahkuch Bala U: Khanabad U: Gahkuch Khari (c)	6.50

(*Continued*)

Table 10.1 Continued

District	*Natural hazards*	*Sample sites: control (c), rural (R), urban (U)*	*Total hazard score*
Gilgit	Floods, landslides, riverbank erosion, avalanches, earthquakes	R: Rahimabad (M'dass) R: Jutal (c) U: Aminabad U: Astore Colony U: Diamer Colony U: Noor Colony Urban U: Noorabad Extension U: Sakarkoi U: Soni Kot U: Wahdat Colony U: Yasin Colony U: Zulfiqarabad U: Jagir Baseen (c) U: Jutial Kot (c) U: Konodas (c) U: Sakwar (c)	6.85
Hunza	Glacier-induced flash floods, GLOFs, earthquakes, landslides	R: Kirmin U: Aliabad Centre U: Hassan Abad Aliabad U: Rahimabad Aliabad R: Hussainabad (c)	5.38
Kharmang	Landslides, earthquakes, floods	None	2.75
Nagar	Landslides, earthquakes, floods	R: Broshal U: Sikandarabad (c)	5.68
Shigar	Avalanches, GLOFs, landslides; glacier-induced flash floods, earthquakes	R: Duskhore Hashupi	6.73
Skardu	Glacier-induced flash floods, earthquakes	R: Chandupa R: Thorgu Paeen (c)	4.98

Notes: Landslides includes rockslides, avalanches, mudslides (Hussain et al. 2022); GLOF: glacial lake outburst flood
Source: Total hazard score from Baig et al. 2021.

The impact of these natural hazards on communities can be measured in terms of vulnerability or 'the way a hazard will affect human lives and properties' by taking into account factors such as literacy, income, and poverty levels, with child dependency, poverty, and topography considered 'the root causes of vulnerability of mountain communities ... followed by those having little agricultural land, lower income opportunities' (Baig et al. 2021: 2009). The total risk to communities can then be derived by multiplying the total hazard score with the total vulnerability score (Table 10.2). Total risk is highest in Shigar (44.4), Diamer (42.7), Ghanche (36.3), and Astore (30.1) where total hazard scores are medium to high, and the income range is low and hence total vulnerability high (Table 10.2). Conversely, total risk is lowest in Hunza (16.4), Kharmang (20.4), and Skardu (22.6) even where total hazard is moderate because incomes are high (e.g. Hunza) or where incomes are low

Table 10.2 Gilgit-Baltistan: Risk and vulnerability indicators and scores, district, 2021

District	*Multi-hazard score**	*Total hazard score*	*Literacy rate (%)*	*Income range*	*Poverty rate (%)*	*Total vulnerability score*	*Total risk score*
Astore	High	4.78	55.1	Low	28.5	6.3	30.1
Diamer	High	6.48	66	Low	57.2	6.6	42.7
Ghanche	High	4.90	43.6	Low	19.5	7.4	36.3
Ghizer	High	6.50	64	Average	11.5	4	26.0
Gilgit	Extreme	6.85	67.1	High	23.5	4.1	27.7
Hunza	High	5.38	71.8	High	13.7	3.1	16.4
Kharmang	High	2.75	49.9	Low	27.6	7.4	20.4
Nagar	High	5.68	66.4	Average	13.7	5.1	28.7
Shigar	High	6.73	46.4	Low	25.4	6.6	44.4
Skardu	High	4.98	53.6	Average	21.3	4.6	22.6

Note: Total risk score = total hazard score x total vulnerability score. Figures have been rounded up.
Source: Baig et al. 2021; *Hussain et al. 2022.

but total hazard is also low (e.g. Kharmang) (Table 10.2). Total risk can thus serve as a lens for understanding the threat of natural hazards and the ability of communities to cope and restore water infrastructure damaged by floods, landslides, and other natural hazards in the region.

10.3 WASEP natural hazard mitigation

A key feature of WASEP is its strong emphasis on engineering, infrastructure, and water quality, and an important part of this is a mandatory multi-hazard vulnerability and risk assessment (MHVRA) at the project identification, planning, and implementation phases, with AKAH taking the lead as the implementing NGO. WASEP MHVRA reports are prepared by a team of geologists, engineers, and GIS specialists from AKAH with feedback from the community. The team evaluates natural hazards in the drinking water source area, along the water supply line, intake chambers, and water tanks. The MHVRA helps AKAH identify safe sites, potential hazards, and mitigation measures. WSC members are then trained to deal with natural hazards and maintenance to maintain system functionality.

Engineering solutions include the careful location and protection of the water intake and water supply line routes, the use of durable materials, and laying pipe networks 4 ft (1.2 m) underground to prevent freezing and protect against natural hazards such as landslides. The storage tank and distribution chamber are located following a hazard risk assessment and protected from surface runoff (AKPBS-P 2012). The locations for civil structures (e.g. intake chamber, reinforced concrete storage tank, sedimentation tank, upflow roughing filters, slow sand filters, valve box, distribution chamber, sump) are

Table 10.3 Engineering audit: Overview, WASEP and control sites, 2021

Settlement: rural (R), urban (U)	*Operational years to 2021 (avg)*	*Excavation depth (m, avg)*	*Pipe length (km, avg)*	*Infrastructure score (1–5): intake, route, network (avg)*	*Safe from natural hazards (intake, supply route, %, avg)*	*Disruption from natural hazards (days: avg, median)*
R: WASEP	12.3	1.2	8.9	3.6	70.8	276.4, 5
U: WASEP	6.5	1.0	12.8	4.8	90	188.3, 0
R: Control	11.2	0.4	9.0	2.4	70	20, 15
U: Control	34.5	0.2	11.2	1.9	70	28, 25

based on a hazard risk assessment and protected from natural hazards through careful site selection, with remedial protective measures taken if required.

The difference in WASEP engineering design is reflected in the different engineering audit scores for WASEP and control sites overall. Table 10.3 provides an overview of the engineering features of WASEP and control sites, and how these are reflected in the infrastructure scores following an engineering audit conducted by the research team. As the first rural WASEP project in the sample was introduced in 1998 (in Broshal, district Nagar), the average age of rural WASEP schemes (12.3 years to 2021) is comparable to rural control (public) schemes (11.2 years). Urban control schemes in contrast are much older (average 34.5 years) as urban water services were introduced much earlier by the government and much later by WASEP (average 6.5 years). This age difference partly explains the lower average infrastructure scores for urban control schemes (1.9 out of 5) compared to both urban WASEP (4.8) and rural control (2.4).

Nonetheless, higher scores for both rural and urban WASEP schemes also reflect some of the engineering features described above. The average excavation depth of the pipe network for example is 4.0 ft (1.2 m) (rural) and 3.2 ft (1.0 m) (urban) for WASEP schemes compared with 1.3 ft (37 cm) (rural) and 0.7 ft (21 cm) (urban) for control schemes. It is worth noting here that for control sites, these averages conceal a range of excavation depths between 0 ft (i.e. pipes laid on the surface of the ground) and 2 ft underground. Similarly, the percentage of schemes with safe water intake (source) and supply routes is higher in WASEP sites (70.8% rural, 90% urban) than control sites (70% rural, 70% urban). The last column in Table 10.3 provides the average and median number of days of disruption to water services as a result of natural hazards for WASEP and control sites. The large difference between the average number of days disrupted in WASEP schemes (276 days rural, 188.3 days urban) and the median number of days disrupted (5 days rural, 0 days urban) indicates that the number of days disrupted have been skewed by a few sites which are examined in further detail in Table 10.4. The median days lost to disruption are lower than control sites (15 days rural, 25 days urban).

Table 10.4 Engineering audit: Infrastructure features and natural hazard impact, rural WASEP and control sites by district

District	*Total risk index*	*Settlement*	*Excavation depth, m*	*Pipe length, km*	*Infrastructure score, avg*	*Natural hazard disruption, days*	*Cause of disruption*	*Damage*
Astore	30.1	Dushkin	1.2	18.5	4.7	3	Floods (2010)	Water intake chamber
Diamer	42.7	Shilati	1.2	12.3	5.0	1	Floods (2010)	Water intake chamber, supply line
Ghizer	26.0	Daeen Chota	1.2	7.0	4.0	3	Floods (2010)	Pipes
		Hatoon Paeen	1.2	19.2	4.0	4	Floods (2010)	Water tank
		Nazirabad	1.2	6.5	5.0	6	Landslide (2012)	Water intake chamber
		Shamaran Paeen	1.2	4.7	4.5	60	Floods (2010)	Water intake chamber
		Singul Shyodass	1.2	1.0	4.7	2	Floods (2010)	Water intake chamber
Gilgit	27.7	Rahimabad (Matumdass)	1.2	8.3	4.0	1	Floods (2010)	Water intake chamber
Hunza	16.4	Kirmin	1.2	9.1	2.7	300	Avalanche (seasonal)	Water tank, pipes
Nagar	28.7	Broshal	1.2	7.3	3.7	7	Slope instability, land drift	Water tank
Shigar	44.4	Duskhore Hashupi	1.2	3.0	5.0	10	Floods (2010)	Water tank
Skardu	22.6	Chandupa	1.2	9.9	2.7	2,920	Floods (2010)	Water tank, main pipeline
Astore	30.1	Doyan (control)	0.15	7.0	2	0	N/A	N/A
Diamer	42.7	Samigal (control)	0.3	8.0	2.3	0	N/A	N/A
Ghizer	26.0	Damas (control)	0.3	12.0	2.7	10	Floods (2010)	Water intake chamber
Gilgit	27.7	Jutal (control)	0.3	13.0	2.3	30	Floods (2010)	Water channel
Hunza	16.4	Hussainabad (control)	0.9		3	60	Floods (2010)	Water intake chamber, supply line
Skardu	22.6	Thorgu Paeen (control)	0 3	5.0	2.3	20	Floods (2010)	Water intake chamber

The main difference between rural and urban WASEP schemes is in the number of sites affected by natural hazards, with all 12 rural WASEP schemes reporting disruptions (Tables 10.4 and 10.5), nine (75%) of which were due to floods. In comparison only four (25%) of the urban WASEP schemes were disrupted by natural hazards although these projects were much more recent (average 6.5 compared with 12.3 operational years for rural schemes). Age may have been a factor as four (67%) of the urban control sites were disrupted although this is the same figure for rural control sites. As mentioned earlier, the higher average number of days disrupted for WASEP schemes has been skewed by two rural schemes (Kirmin, 300 days and Chandupa, 2,920 days) and two urban schemes that shared the same water source that was damaged (Aliabad Centre, 1,500 days and Rahimabad Aliabad, 1,500 days).

Damage to water infrastructure in three of these sites was in part related to poor engineering design. In the case of Kirmin, the WSC had pressured AKAH engineers to change the design of the water system in order to reduce the community's labour contribution to the project. This compromised the design, with the new location for the water tank and route of the pipe supply line exposed to avalanches. Aliabad Centre was unusual in that water infrastructure was provided by both the Public Works Department (PWD) (feeder pipeline and water storage tank) and AKAH (water supply network in town). Disruption was due to damage to the feeder pipeline laid on the surface of the ground with almost all of this exposed to landslides and rockfalls. The water intake was damaged by movement of the Shishper glacier in 2019 (see Shah et al. 2019) which also impacted the town of Rahimabad Aliabad, which shared the same water source. While the water system was not completely non-functional, a large part was damaged and the disruption of 1,500 days for each site here drastically skews the average number of days lost for urban WASEP schemes that only had two other episodes of disruption (Chokoporo 1 day, Hassan Abad Aliabad 12 days) (Table 10.5).

Disruption as a result of these natural disasters was partly reflected in lower average engineering audit scores for Aliabad Centre (2 intake, 5 route, 5 network, and 4 average), Kirmin (3 intake, 1 route, 4 network, and 2.7 average), and Chandupa (3 intake, 2 route, 3 network, and 2.7 average) (Tables 10.4 and 10.5). However, AKAH was not responsible for the source (intake) and route in Aliabad Centre, and the source and route in Kirmin was not consistent with WASEP SOPs. Conversely, disruptions despite high engineering scores, for example in Shamaran Paeen (60 days but 4.5 average infrastructure score) and Duskhore Hashupi (10 days and 5 average infrastructure score), suggest that there are limits to engineering design in the context of high risks posed by natural hazards. Moreover, the community's capacity to respond to damage arising from natural hazards can play an important role in determining the impact of natural hazards in terms of restoring any disrupted water services.

Table 10.5 Engineering audit: Infrastructure features and natural hazard impact, urban WASEP and control sites by district

District	*Total risk index*	*Settlement*	*Excavation depth, m*	*Pipe length, km*	*Infrastructure score, avg*	*Natural hazard disruption, days*	*Cause of disruption*	*Damage*
Ghizer	26.0	Chokoporo Gahkuch Bala	1.2	18.9	4.7	1	Rockfall (2018)	Water source
		Domial Buridur Gahkuch Bala	1.2	6.1	4.7	0	N/A	N/A
		Khanabad	1.2	6.1	4.7	0	N/A	N/A
Gilgit	27.7	Aminabad	0.9	15.5	5.0	0	N/A	N/A
		Astore Colony	0.9	6.1	5.0	0	N/A	N/A
		Diamer Colony	0.9	2.9	5.0	0	N/A	N/A
		Noor Colony	0.9	7.9	5.0	0	N/A	N/A
		Noorabad Extension	0.9	18.6	5.0	0	N/A	N/A
		Sakarkoi	0.9	7.9	5.0	0	N/A	N/A
		Soni Kot	0.9	16.4	5.0	0	N/A	N/A
		Wahdat Colony	0.9	17.7	5.0	0	N/A	N/A
		Yasin Colony	0.9	13.9	5.0	0	N/A	N/A
		Zulfiqarabad	0.9	25.2	5.0	0	N/A	N/A
Hunza	16.4	Aliabad Centre	1.2	14.0	4.0	1,500	Glacial movement (2019), GLOF, rockfall	Water intake chamber, supply line
		Hassan Abad Aliabad	1.2	12.1	4.3	12	Rockfall	Supply line
		Rahimabad Aliabad	1.2	21.4	4.0	1,500	Glacial movement (2019), GLOF, rockfall	Water intake chamber, supply line

(*Continued*)

Table 10.5 Continued

District	*Total risk index*	*Settlement*	*Excavation depth, m*	*Pipe length, km*	*Infrastructure score, avg*	*Natural hazard disruption, days*	*Cause of disruption*	*Damage*
Ghizer	26.0	Gahkuch Khari (control)	0.3	N/A	N/A	0	N/A	N/A
Gilgit	27.7	Jagir Baseen (control)	0.15	13.0	2.7	0	N/A	N/A
		Jutial Kot (control)	0.15	8.0	1.7	20	Flash floods (seasonal)	Water intake chamber
		Konodas (control)	0.15	12.0	2.0	60	Flash floods (seasonal)	Water intake chamber
		Sakwar (control)	0.15	11.0	2.0	30	Flash floods (seasonal)	Water channel
Nagar	28.7	Sikandarabad (control)	0.3	12.0	1.3	60	Flash floods (seasonal), landslides	Water intake chamber, supply line

Note: GLOF, glacial lake outburst flood

10.4 Community responses to natural hazards

The sustainability of water infrastructure and services depends not only on the risks posed by natural hazards (as captured by the multi-hazard and total hazard scores) but also the vulnerability of affected communities (captured by the total vulnerability score that includes education and income levels and that indicates the coping capacity of communities). Together hazard and vulnerability can be captured in a total risk index that has been calculated for each of the 10 districts in Gilgit-Baltistan (Baig et al. 2021) (Table 10.2). Tables 10.6 and 10.7 summarize the impact of natural hazards on water infrastructure and the last three columns add information on whether repairs were undertaken, the role of the community, and whether external support was provided.

Overall, 24 (60%) WASEP and control sites were disrupted by natural hazards. Of these 24 water schemes, only Chandupa (WASEP) did not undertake repairs or seek external help, with three control sites (Damas, Jutial Kot, Konodas) getting external support from the Local Government & Rural Development Department (LG&RDD) and PWD. Eight (66.7%) rural and all four (100%) urban WASEP WSCs received external support for repairs to restore water infrastructure, with AKAH providing support to five (62.5%) rural communities, and local government departments (PWD, LG&RDD) supporting three (37.5%) rural and all four (100%) urban WSCs. Eleven (91.7%) rural and all four (100%) urban WASEP WSCs contributed labour compared to three (75%) rural and two (50%) urban control sites. Additionally, eight (66.7%) rural WASEP communities contributed cash on top of labour, with two (50%) urban WASEP WSCs drawing from the WSC fund. This appears to be consistent with rural WSCs tending to only collect tariffs or financial contributions as and when needed, as opposed to collecting regular (monthly) tariffs (see Chapter 8).

Vulnerability scores can help identify which communities are most at risk in the event of a natural hazard. They can also provide some indication of the coping capacity of a community and its ability to respond to damaged water infrastructure; the total hazard and total vulnerability scores and total risk scores for each district (Table 10.2) allow for a deeper investigation into how some of these factors (specifically literacy and income) help communities mobilize and undertake repairs. Tables 10.8 and 10.9 supplement the selected vulnerability scores (literacy, income, poverty) on the district level (Baig et al. 2021) with proxies (highest educational attainment and primary occupation) from the household survey of the WASEP engineering sub-sample.

The evidence suggests that rural WASEP communities in the engineering sub-sample are much better educated than the district-level literacy rate might indicate; seven (58.3%) sites have either a master or graduate degree as the highest educational attainment and another six (50%) have intermediate or matriculation qualifications (Table 10.8). Urban figures are even higher: all four (100%) WASEP sites have graduate and/or master level degrees

Table 10.6 Community responses to natural hazards, rural WASEP and control

District	*Total risk index*	*Settlement*	*Infrastructure score (avg)*	*Disruption (days)*	*Cause*	*Damage*	*Repairs*	*Community contribution*	*External support*
Astore	30.1	Dushkin	4.7	3	Floods (2010)	Water intake chamber	Yes	Labour, cash	No
Diamer	42.7	Shilati	5.0	1	Floods (2010)	Water intake chamber, supply line	Yes	Labour, cash	AKAH
Ghizer	26.0	Daeen Chota	4.0	3	Floods (2010)	Pipes	Yes	Labour, tariffs, cash	AKAH
		Hatoon Paeen	4.0	4	Floods (2010)	Water tank	Yes	Labour, cash	PWD
		Nazirabad	5.0	6	Landslide (2012)	Water intake chamber	Yes	Labour	AKAH
		Shamaran Paeen	4.5	60	Floods (2010)	Water intake chamber	Yes	Labour, cash	AKAH
		Singul Shyodass	4.7	2	Floods (2010)	Water intake chamber	Yes	Labour, cash	No
Gilgit	27.7	Rahimabad (Matumdass)	4.0	1	Floods (2010)	Water intake chamber	Yes	Labour, cash	No
Hunza	16.4	Kirmin	2.7	300	Avalanches	Water tank, pipes	Yes	Labour	AKAH
Nagar	28.7	Broshal	3.7	7	Slope instability, land drift	Water tank	Yes	Labour, cash	LG&RDD
Shigar	44.4	Duskhore Hashupi	5.0	10	Floods (2010)	Water tank	Yes	Labour	LG&RDD
Skardu	22.6	Chandupa	2.7	2,920	Floods (2010)	Water tank, main pipeline	No	No	No
Ghizer	26.0	Damas (control)	2.7	10	Floods (2010)	Water intake chamber	Yes	No	LG&RDD
Gilgit	27.7	Jutal (control)	2.3	30	Floods (2010)	Water channel	Yes	Labour	No
Hunza	16.4	Hussainabad (control)	3.0	60	Floods (2010)	Water intake chamber, supply line	Yes	Labour	No
Skardu	22.6	Thorgu Paeen (control)	2.3	20	Floods (2010)	Water intake chamber	Yes	Labour	No

Notes: AKAH, Aga Khan Agency for Habitat; PWD, Public Works Department; LG&RDD, Local Government & Rural Development Department.

Table 10.7 Community responses to natural hazards, urban WASEP and control

District	*Total risk index*	*Settlement*	*Infrastructure score (avg)*	*Disruption (days)*	*Cause*	*Damage*	*Repairs*	*Community contribution*	*External support*
Ghizer	26.0	Chokoporo Gahkuch Bala	4.7	1	Rockfall (2018)	Water source	Yes	Labour	LG&RDD
Hunza	16.4	Aliabad Centre	4.0	1,500	Glacial movement (2019), GLOF, rockfall	Water intake chamber, supply line	Yes	WSC fund, labour	PWD
		Hassan Abad Aliabad	4.3	12	Rockfall	Supply line	Yes	Labour	LG&RDD
		Rahimabad Aliabad	4.0	1,500	Glacial movement (2019), GLOF, rockfall	Water intake chamber, supply line	Yes	WSC fund, labour	PWD
Gilgit	27.7	Jutial Kot (control)	1.7	20	Flash floods (seasonal)	Water intake chamber	Yes	No	PWD
		Konodas (control)	2.0	60	Flash floods (seasonal)	Water intake chamber	Yes	No	PWD
		Sakwar (control)	2.0	30	Flash floods (seasonal)	Water channel	Yes	Labour	*External support**
Nagar	28.7	Sikandarabad (control)	1.3	60	Flash floods (seasonal). Landslides	Water intake chamber, supply line	Yes	Labour	No

Notes: *No information on actual external support provided; GLOF, glacial lake outburst flood; LG&RDD, Local Government & Rural Development Department; PWD, Public Works Department.

Table 10.8 Vulnerability indicators and community contribution, rural WASEP

District	*Literacy rate (%)*	*Income range*	*Poverty rate (%)*	*Settlement*	*Highest educational attainment*	*Primary occupation*	*Repairs*	*Community contribution*	*External support*
Astore	55.1	Low	28.5	Dushkin	22% master's 20% graduate	30% government office 15% agriculture own land	Yes	Labour, cash	No
Diamer	66.0	Low	57.2	Shilati	19% intermediate 19% middle	27% agriculture labour 23% construction labour	Yes	Labour, cash	AKAH
Ghizer	64.0	Average	11.5	Daeen Chota	32% intermediate 23% matriculation	20% military 16% own business 16% trade	Yes	Labour, tariffs, cash	AKAH
				Hatoon Paeen	24% intermediate 23% matriculation	15% military 13% construction labour	Yes	Labour, cash	PWD
				Nazirabad	31% intermediate 31% matriculation	28% military 17% agriculture own land	Yes	Labour	AKAH
				Shamaran Paeen	45% matriculation 27% intermediate	47% military 13% own business	Yes	Labour, cash	AKAH
				Singul Shyodass	32% graduate 22% master's	36% government office 14% construction labour	Yes	Labour, cash	No
Gilgit	67.1	High	23.5	Rahimabad (Matumdass)	23% matriculation 21% graduate 20% master's	26% government office 16% construction labour	Yes	Labour, cash	No
Hunza	71.8	High	13.7	Kirmin	67% master's 17% intermediate	24% agriculture land 20% construction labour	Yes	Labour	AKAH
Nagar	66.4	Average	13.7	Broshal	28% master's 26% intermediate	22% government office 19% agriculture land	Yes	Labour, cash	LG&RDD
Shigar	46.4	Low	25.4	Duskhore Hashupi	34% graduate 30% intermediate	23% government office 20% government manual	Yes	Labour	LG&RDD
Skardu	53.6	Average	21.3	Chandupa	24% matriculation 18% illiterate	26% agriculture labour 26% agriculture own land	No	No	No

Table 10.9 Vulnerability indicators and community contribution, urban WASEP

District	*Literacy rate (%)*	*Income range*	*Poverty rate (%)*	*Settlement*	*Highest educational attainment*	*Primary occupation*	*Repairs*	*Community contribution*	*External support*
Ghizer	64.0	Average	11.5	Chokoporo Gahkuch Bala	23% matriculation 21% graduate 20% master's	26% government office 16% construction labour	Yes	Labour	LG&RDD
Hunza	71.8	High	13.7	Aliabad Centre	46% master's 30% graduate	47% own business 16% salaried private sector 16% government office	Yes	WSC fund, labour	PWD
				Hassan Abad Aliabad	28% master's 21% intermediate	36% government office 13% other	Yes	Labour	LG&RDD
				Rahimabad Aliabad	40% master's 27% graduate	35% own business 19% salaried private sector	Yes	WSC fund, labour	PWD

(Table 10.9). Similarly, the primary household occupation does not appear to correspond with the district income indicator, with sites in high-income districts (e.g. Gilgit and Hunza) featuring manual work (e.g. construction labour) as the primary occupation. The community's cash contribution to repairs also does not appear to be related to the primary occupation in each site; sites with manual labour as the primary occupation (e.g. Shilati and Singul Shyodass) contribute cash for repairs (Table 10.8).

Nonetheless, the community coping capacity as reflected in the highest educational attainment and primary occupation can offer an explanation for the contrasting community responses to disruptions from natural disaster, specifically the role of WSCs. In Aliabad Centre, damaged parts of the supply line were replaced by the community on a self-help basis, led by an active WSC after the lack of support from the district administration. The scheme in Chandupa worked well until floods in 2010 damaged the main storage tanks and parts of the water infrastructure were subsequently stolen by members of the community. The WSC was inactive and did not seek external help, mobilize the community to rehabilitate the scheme, or use available funds for repairs, and the community did not contribute labour or cash. The community reportedly did not know where their endowment fund had been deposited. In Duskhore Hashupi an active WSC with the help of the community rebuilt the project and undertook regular maintenance. The WSC also identified a new source of cleaner water, collected contributions from households, government agencies, and political leaders, and invested PKR5 m (US$48,663 based on exchange rates for 2009–2019) on pipes to connect to the new source.

The performance of both WSCs is also consistent with the household survey results and proxy vulnerability indicators. In Aliabad Centre, over 70% of respondents rated the WSC 'OK' or 'good', 76% of households had a master's or graduate degree, 47% owned their business, and 32% were salaried private sector or white-collar government employees. In Chandupa, in comparison, around 30% of households rated the WSC 'OK' and over 40% 'bad' or 'very bad', 18% of households were illiterate, and 26% worked as agriculture labour (Tables 10.8 and 10.9).

10.5 Conclusion

The evidence on the sustainability of WASEP in the context of natural hazards suggests that engineering design is important but, while this may mitigate the impact of natural hazards, it cannot always prevent damage to water infrastructure. Nonetheless, better engineering design is reflected in higher infrastructure scores in WASEP compared to control sites, with a lower median number of days lost from disruptions caused by natural hazards in the rural sample which is broadly comparable in terms of operational years and pipe length (Table 10.3). The urban WASEP sample is too recent (6.5 years average) for meaningful comparisons with urban control schemes (34.5 years average) but nonetheless scores much higher in terms of infrastructure design (Table 10.3).

Just as important is the capacity of WSCs and communities to cope and undertake repairs to damaged water infrastructure. The WASEP scheme with the greatest number of days lost to natural hazards is Chandupa (2,920 days), which was also the only non-functional scheme in the engineering sub-sample with an inactive WSC that failed to undertake repairs or seek external support. In contrast, other WSCs were able to respond by mobilizing community labour or cash contributions, with over half of rural WSCs and all urban WSCs seeking and receiving external support from AKAH or government agencies. The failure of the WSC in Chandupa may be a reflection of vulnerability indicators, in this case educational attainment and primary household occupation, but there is no clear indication if these were factors in the responses of other WSCs in their ability to undertake repairs (Tables 10.8 and 10.9). The fact that none of the control sites contributed cash suggests a greater willingness and perhaps greater sense of ownership among WASEP communities. The evidence also suggests that external support is important to restore water services after damage from natural hazards.

CHAPTER 11

How sustainable and scalable is WASEP?

Jeff Tan

Community-based water management (CBWM) continues to be the leading model for the delivery of (rural) drinking water supply schemes (DWSS) despite the lack of evidence of increased functionality (see e.g. Broek 2009). In fact, early evidence has already shown that community management 'has been no more successful in delivering a sustainable water supply than any other approach' (Schouten and Moriarty 2003: 1). Underlying this has been the lack of financial and management capacities, particularly of poor (rural) communities, to maintain operations (see e.g. McCommon et al. 1990; World Bank 2017) along with the absence of long-term financing not only for capacity building but also for major repairs and system rehabilitation and service expansion.

This lack of financing for operating expenses (OpEx) associated with operations and maintenance (O&M) and capital maintenance expenditure for major repairs and system rehabilitation is not restricted to the CBWM model. After all, CBWM emerged as an alternative to public water provision because local authorities typically did not have the resources to maintain or rehabilitate existing water services let alone extend services to unserved populations. However, both CBWM and direct local government provision are the least sustainable models in terms of financing water systems, with '(a)sset management of small water schemes, managed by communities or local government ... mostly absent' (World Bank 2017: 47). This is because rural water services typically lack clarity around asset ownership and the designation of responsibilities for capital maintenance (including minor versus major repairs) and asset renewal, with the result often being no differentiation between operational, capital maintenance, and capital replacement costs and hence no sustainable financing mechanisms for these (World Bank 2017).

The problem with CBWM is that it does not resolve this funding gap and instead requires additional resources to build community capacities. Rather than re-evaluating the appropriateness of community management in light of these two constraints in the CBWM model, low levels of functionality and the lack of sustainability have been, and continue to be, blamed on the absence of an enabling environment, specifically policy, institutional, technical, and financial support to strengthen community management capacities. But this external support was always a key condition for sustainable CBWM (see e.g. McCommon et al. 1990), which raises the question why this support has not been provided (see e.g. World Bank 2017). This can only be understood in terms of the inherent limitations of the CBWM model and wider institutional context. Specifically, a fragmented financing model that favours short-term project and programme funding over long-term sector-wide investment and the lack of state institutional capacity to support and finance CBWM has precluded long-term support for communities and financing to maintain water infrastructure. Additionally, the lack of financial capacity among many communities has resulted in irregular and non-payment of tariffs that is compounded by very low tariffs (to ensure affordability) that undermine cost recovery.

There is growing recognition on the policy level of these financing constraints, with proposals including 'water financing facilities' to tap into private financing (see e.g. Fonseca 2015) and the professionalization of community management (World Bank 2017). Unfortunately, neither of these fully addresses the problem of financing and specifically, the ongoing and increasing need for public funds. On the ground, senior management at the Aga Khan Agency for Habitat (AKAH), the implementing agency of the Water and Sanitation Extension Programme (WASEP), are aware of the dependency on external donors and need for budgetary allocations in the government's Annual Development Program along with a government endowment fund to supplement community funds for service expansion. With adequate financing, sustainability and scaling up are possible for CBWM although this would also apply to other models of service provision. In a recent study of rural water delivery models in 16 developing countries, the CBWM model scored poorly for financing and asset management for point sources, slightly better for piped sources, and higher for federations of service providers such as farmers cooperatives, with public utility provision exhibiting 'the best conditions for sustainability' (World Bank 2017: 67).

The chapters in this edited volume examined how these problems with the CBWM model are reflected in the case study of WASEP, and how these have ultimately constrained its sustainability. WASEP is especially instructive as a successful implementation of CBWM, not just in terms of its coverage but also in its engineering complexity. The success of WASEP can be defined and assessed in terms of: (a) the functionality of schemes; (b) the ability of water and sanitation committees (WSCs) to undertake O&M; (c) water infrastructure and water quality; (d) sustained operations; and (e) indefinite operations. By most measures, WASEP must be considered a successful implementation of CBWM. Since its introduction in 1997, it has delivered clean drinking water through piped water systems (not handpumps) to 459 settlements covering 47,629 households with an estimated population of over 400,000 people. Evidence of sustainability and scalability is drawn from responses to a household survey of 3,132 households based on a random sample of 25 rural and 25 urban WASEP projects that covered 13,694 households and 112,653 people (Chapter 2). This was supplemented by WSC interviews, focus group discussions, engineering audits, and water quality tests.

WASEP has been successful in terms of (a) functionality, with only one (rural) scheme from the sample not functional and almost all WSCs active at the point of inspection and for more than six months until December 2021. WASEP schemes performed significantly better than control sites for infrastructure quality and water quality, with most WASEP schemes complying with WHO criteria for safe drinking water compared to a fraction of rural control sites and no urban control sites (Chapter 9). More crucially, WASEP schemes were also able to (d) sustain operations; 36% of the rural sample exceeded the 15-year projected lifespan of water infrastructure with an average operating age of 13.6 years for the rural sample. Urban projects were introduced too recently to assess sustained operations, with an average operating age of 4.8 years. The continued functioning of rural WASEP projects beyond their natural lifespan is itself testament to the quality of engineering and water infrastructure, and resilience of communities. However, this also highlights the problem of financing for rehabilitation and service expansion to meet the needs of growing populations, particularly in the context of rapid urbanization.

The functionality and sustained operations are especially remarkable given the low levels of tariffs and non-collection/non-payment of regular tariffs. Although 68% of rural WSCs collected tariffs, 28% only collected this as and when repairs were needed, and only 39% of rural households were willing to pay regular tariffs despite high rates of rural participation and low tariff levels (Chapter 8). Rural communities have thus been able to sustain operations without regular tariff payments. This raises the question of whether participation matters for sustainability given that high levels of rural participation have not translated into greater willingness or ability to pay regular tariffs. It should be noted that the ability of communities to raise funds for repairs is not the same as preventive maintenance that requires a water and sanitation operator or plumber whose salary depends on the monthly payment of tariffs. In other words, while (minor) repairs may have been undertaken to sustain operations, this is not the same as maintenance. This suggests that engineering design and infrastructure quality has played a part in the continuous functioning of rural WASEP schemes, despite the absence of preventive maintenance, as reflected in the water quality test results and higher water quantity/quality (Chapters 9 and 10), and user satisfaction in the household survey.

However, the tension between affordability and cost recovery that characterizes CBWM is also reflected in WASEP: tariffs were set below recommendations by AKAH, and cost recovery was compounded by irregular tariff collections and payments, especially in poorer rural communities and among poorer rural households (Chapter 8). As a result, (b) O&M performance is mixed. While most rural and all urban WSCs were functioning and had (mandatory) O&M (endowment) funds in place, low levels of tariff collection meant that 76% of rural WSCs were estimated to have monthly operating deficits and 67% had drawn down their O&M funds, recording a lower closing balance than at the start of the project, presumably to pay for maintenance and repairs. Only 37.5% of rural WSCs interviewed reported being able to cover both OpEx and the costs of minor repairs although five of these were estimated to have monthly operating deficits.

Urban households are more willing to pay higher and more regular tariffs, and almost all urban WSCs in the sample had regular tariff collections in place. As a result, 62.5% of urban WSCs are estimated to have operating surpluses, and 69% have increased the value of their O&M funds. This is not surprising given the wider socioeconomic context and specifically higher household incomes in urban communities, despite the higher capital and operating costs. Paradoxically, CBWM emerged as a sustainable alternative to deliver or maintain water services in relatively small, isolated, and sparsely populated rural communities (McCommon et al. 1990), but these communities have always lacked the necessary capacities and support; external support and an enabling environment have still not been provided more than 30 years after the need was identified. And while urban communities may have better management and financial capacities, CBWM was never originally envisaged for densely populated urban centres where centralized (and publicly provided and managed) water services have made more sense.

It is worth noting that the successful delivery and operations of WASEP also lie in the careful screening by AKAH of partnering communities. The terms of partnership seek to balance community needs with community capacity to finance and manage water services but also mean that only communities that demonstrate the willingness and ability to manage get selected. This necessarily excludes communities lacking in management capacity, village organizational structures, and prior experience in management and working with NGOs, which means that the WASEP model (and CBWM by extension) cannot be seen as a solution for universal water provision if not all communities qualify. Even then, the selected rural and urban WSCs in the research sample were unable to meet the costs of major repairs and capital maintenance expenditure associated with system rehabilitation.

Most WSCs needing major repairs requested external support but this cost component had not been budgeted for and it is not clear who should finance this. While the project is handed over to communities upon completion, ownership of actual water assets or infrastructure remains unclear, with the state being the ultimate owner especially in the absence of legal ownership by communities. However, legal ownership as a solution recommended in the literature cannot resolve the issue of long-term financing and, if anything, leaves communities even more exposed as the state would then no longer have any legal obligation to provide support. As it stands, this legal ambiguity means that communities are expected to raise their own contributions or seek external funding for major repairs, with government departments stepping in on an ad hoc basis.

The final measure of sustainability is (e) indefinite operations or, more practically, the conditions that can sustain indefinite operations. As mentioned above, the CBWM model of project-based financing means that the community must seek external funding at the end of the natural lifespan of water infrastructure for asset renewal (system rehabilitation). Although there is separate funding under WASEP's 'rehabilitation only' projects, this covers only a small fraction of total WASEP schemes in Gilgit-Baltistan as the priority is to extend coverage to previously unserved communities. The trade-off is water systems that are unsustainable because the community cannot be expected to undertake system rehabilitation let alone system expansion to ensure water services keep up with the demands of growing (urban) populations.

Conversely, longer-term financing for capacity building and other external support required for sustainability necessarily means that fewer communities benefit from immediate access to clean water. Even then, this does not address the problem of financing indefinite operations 'for the lifetime of a community' (Schouten and Moriarty 2003: ix). Nor will it facilitate a wider enabling environment as this would require financing state capacities to provide the necessary policy and institutional support. The discussion of sustainability thus requires a wholesale re-examination of the CBWM financing model as financing for indefinite operations will invariably have to come from government, given that donors are unwilling or unable to commit to long-term, let alone ongoing (indefinite), financing.

Rather than address government institutional, capacity, and fiscal constraints, the emphasis in the literature has been on public–private partnerships (PPP), professionalization, and 'innovative financing models' (see e.g. Lockwood and Smits 2011; Hutchings et al. 2015; World Bank 2017; UN Water n.d.). However, the reliance on private sector financing including proposals for 'water financing facilities' downplays the substantial financing requirements of water infrastructure (that accounts for 75% of total expenditure) and the lack of commercial viability that has necessitated significant and ongoing external support, with the bulk of capital costs covered by government (69%) compared to donors (20%) and communities (5–10%) (Carter 2021).

Given the high capital costs and absence of cost-covering tariffs and full-cost recovery, private sector participation through private financing or PPP would require public subsidies because there are no profit opportunities and hence no incentives otherwise for private participation (see e.g. Tan 2008, 2011). PPP is promoted on the basis of private sector efficiency and lower costs, which are not supported by the wider evidence, and despite the fact that public sector borrowing is cheaper (Tan 2011). The alternative is for governments to subsidize communities or to provide services directly. The latter would require a major policy rethink and a return of government to its traditional role as provider rather than as facilitator or enabler of private provision.

The other consideration for PPP, especially in the context of Gilgit-Baltistan, is the absence of the private sector, which is not unexpected given the lack of profit opportunities and public subsidies. As a result, the 'private' in PPP has been redefined and extended to cover anything not 'public', which now includes communities and NGOs that would normally come under the category of civil society. This is highly misleading and inappropriate because communities and NGOs are fundamentally different from the private sector for the simple fact that they are not driven by profit and do not have to pay shareholders. The continued promotion of PPP in this context gives the impression that there is a private sector when the reality is that WASEP and CBWM emerged in the wider context of both state failure (the inability of local authorities to deliver or maintain water services) and market failure (the unwillingness of the private sector to step in). AKAH is a non-profit NGO which is precisely why the WASEP model is so cost-effective and attractive to the provincial Government of Gilgit-Baltistan that funded the two largest WASEP urban schemes in Jutial and Danyore in Gilgit City. The cost of WASEP was reportedly significantly lower than a competing proposal from the Public Works Department that would presumably have necessitated sub-contracting out to private contractors.

The experiences of WASEP raise questions about the CBWM principles of participation, ownership and control, and cost sharing, especially given the importance of the wider social context. Rural households have higher levels of participation and a greater sense of ownership than urban households, but these do not translate into a greater willingness to share project costs. The rural share for capital costs comes in the form of in-kind (unskilled) labour contributions but rural households are less likely to pay regular tariffs and most rural WSCs do not collect tariffs regularly. In contrast, urban households have lower levels of participation, a lower sense of ownership, and lower levels of unity with a higher incidence of reported conflict of longer duration. Despite this, urban households contribute more per household to capital costs (although this translates into a smaller community share of total project cost because urban projects are more expensive) and are more likely to pay higher tariffs regularly, with most urban WSCs collecting tariffs monthly. What emerges then from the WASEP case study is that urban projects are more likely to be financially sustainable in the medium term, setting aside the issue of the long-term financing of major repairs, system rehabilitation or asset renewal, and service expansion.

Similarly, while the participation of women as the main beneficiaries of clean drinking water is seen as important for maintaining water services, women's participation in WASEP is very low, consistent with the evidence elsewhere (see e.g. Soto et al. 2021; Hannah et al. 2021). As expected, women's participation is lower in more socially conservative rural and urban communities but also lower in urban communities as a whole because women may not feel safe given the greater community diversity in terms of sectarian composition and old and new settlers (Chapter 6). Crucially, there is a weak link between women's participation and sustainability, and the promotion of women's participation as an end in itself will be constrained by wider unequal power relations rooted in patriarchal social structures that are typically manifested in terms of tradition, customs, cultural practices, and social norms. The wider socioeconomic context appears to also be an important consideration in how communities respond to the impact of natural hazards that are a feature of Gilgit-Baltistan. However, while the only WSC that failed to respond (and failed operationally) scored poorly in terms of vulnerability, socioeconomic status appears to not matter as much as external support for the successful undertaking of major repairs (Chapter 10).

This reinforces the importance of external technical, institutional, policy, and financial support for communities that has long been identified in the literature on CBWM but that has not been (adequately) provided. The ongoing need of communities for external support and the lack of this support can only be understood in terms of the inherent limitations of the CBWM model related to the inability and/or unwillingness of poor (rural) households to pay regular tariffs, and the fragmented and short-term financing that characterize CBWM. Successful community management thus depends on external support to build capacity, but this support is not possible with the current model of financing. Additionally, government and state institutions in developing countries such as Pakistan that need to provide this support and create a wider 'enabling environment' are typically fragmented and hollowed out, lacking the necessary capacities to introduce, implement, coordinate, and harmonize policies (Chapter 3).

The problem of inadequate state capacities is central, not just for the short-term support of communities and CBWM, but more crucially for the long-term delivery of sustainable water services. This is because the increasing acknowledgement of the 'endemic problems in the sustainability and scalability of this model are leading many to conclude we have reached the limits of an approach that is too reliant on voluntarism and informality' (Hutchings et al. 2015: 963). The CBWM model is characterized by 'waning community interest and a reluctance to volunteer' (Broek and Brown 2015: 21; Chowns 2019) and is 'unable to provide sustainable water service to rural people' (Lockwood and Smits 2011: 1). However, the shift in thinking towards PPPs, professionalization, and 'innovative financing models' (see e.g. World Bank 2017) continues to downplay the central role of government and the state not just as an enabler and facilitator, but also as a provider of water services historically.

The issue of the sustainability of CBWM is fundamentally then about the sustainable provision of water services, which cannot be discussed without a wholesale reassessment of the role of communities, the private sector, and government that needs to be framed by the features and inherent limitations of CBWM and the private sector. The lack of community management capacity can only be partially addressed through technocratic solutions centred on capacity building. This is because technical capacity itself is insufficient given the limits of voluntarism, where community members are unwilling to participate in management committees, particularly in poorer communities (see e.g. McCommon et al. 1990; Hutchings et al. 2016; World Bank 2017). This is reflected in the WASEP sample where the majority of WSCs rely on volunteers as opposed to elections, and where the turnover of WSC members is very low, meaning that community members are generally not willing to volunteer on WSCs (Chapter 5).

In this context, government and state institutions need to be re-engaged in the discussion of 'scaling up in time' (for improved and expanded services to ensure universal coverage) and 'scaling up in space' (for the delivery of water services indefinitely). Rather than the continued bypassing of government and the state, there needs to be a serious reconsideration of the role of government and the rebuilding of state capacities to finance, legislate, coordinate, and implement universal water services that are sustainably maintained and financed. Scalability will also require the formalization of WSCs, which are currently informal community organizations without legal recognition, but the difficulty in legalization is itself illustrative of the wider institutional context characterized by weak state capacity.

The case study of WASEP illustrates the successful implementation of CBWM but also the constraints to sustainability and scalability inherent in this model of service delivery. Nonetheless, the WASEP model can offer important lessons in terms of the design and implementation of DWSS not just for other CBWM schemes but more crucially for government. WASEP's integrated approach that prioritizes engineering and water quality management suggests that elements (or best practices) of this model can be transferred to government to minimize cost and provide strong engineering solutions as the foundations for more sustainable drinking water systems. However, given that the lower cost of WASEP is grounded in the not-for-profit and development ethos of AKAH as part of the wider Aga Khan Development Network, implementing the WASEP model will also require the review of the current emphasis of national and provincial water policies on community participation and PPPs, and recognition of the limitations of both communities and the private sector.

References

Chapter 1 Introduction: Why community water management?

Baumann, E. (2006) 'Do Operation and Maintenance Pay?', *Waterlines* 25(1): 10–12. https://doi.org/10.3362/0262-8104.2006.033

Carter, R. (2021) *Rural Community Water Supply: Sustainable Services for All*, Rugby: Practical Action Publishing.

Churchill, A. (1987) *Rural Water Supply and Sanitation: Time for a Change*, Washington, DC: World Bank Discussion Paper No. 18. Washington, DC: The World Bank. https://documents.worldbank.org/en/publication/documents-reports/documentdetail/840401468764671363/rural-water-supply-and-sanitation-time-for-a-change (last accessed 28 January 2023).

Fonseca, C., Franceys, R., Batchelor, C., McIntyre, P., Klutse, A., Komives, K., Moriarty, P., Naafs, A., Nyarko, K., Pezon, C., Potter, A., Reddy, R. and Snehalatha, M. (2011) *Life-Cycle Costs Approach: Costing Sustainable Services*, WASHCost Briefing Note 1a, The Hague: IRC International Water and Sanitation Centre. https://www.ircwash.org/resources/briefing-note-1a-life-cycle-costs-approach-costing-sustainable-service (last accessed 28 January 2023).

Hannah, C., Giroux, S., Krell, N., Lopus, S., McCann, L.E., Zimmer, A., Caylor, K.K. and Evans, T.P. (2021) 'Has the Vision of a Gender Quota Rule Been Realized for Community-Based Water Management Committees in Kenya?' *World Development*, 137: 1–13. https://doi.org/10.1016/j.worlddev.2020.105154

Harvey, P. and Reed, R. (2006) 'Community-Managed Water Supplies in Africa: Sustainable or Dispensable?', *Community Development Journal*, 42(3): 365–378. https://doi.org/10.1093/cdj/bsl001

Hope, R., Thomson, P., Koehler, J. and Foster, T. (2020) 'Rethinking the Economics of Rural Water in Africa', *Oxford Review of Economic Policy*, 36(1): 171–190. https://doi.org/10.1093/oxrep/grz036

Hutchings, P., Chan, M.Y., Cuadrado, L., Ezbakhe, F., Mesa, B., Tamekawa, C. and Franceys, R. (2015) 'A Systematic Review of Success Factors in the Community Management of Rural Water Supplies over the Past 30 Years', *Water Policy*, 17(5): 963–983. https://doi.org/10.2166/WP.2015.128

Hutchings, P., Franceys, R., Mekala, S., Smits, S. and James, A.J. (2016) 'Revisiting the History, Concepts and Typologies of Community Management for Rural Drinking Water Supply in India', *International Journal of Water Resources Development*, 33(1): 152–169. https://doi.org/10.1080/07900627.2016.1145576

Kelly, E., Lee, K., Shields, K., Cronk, R., Behnke, N., Klug, T. and Bartram, J. (2017) 'The Role of Social Capital and Sense of Ownership in Rural Community Managed Water Systems: Qualitative Evidence from Ghana, Kenya, and Zambia', *Journal of Rural Studies*, 56: 156–166. https://doi.org/10.1016/j.jrurstud.2017.08.021

Lockwood, H. (2004) 'Scaling Up Community Management of Rural Water Supply', Thematic Overview Paper, IRC International Water and Sanitation Centre. https://www.ircwash.org/sites/default/files/Lockwood-2004-Scaling.pdf (last accessed 14 June 2024).

Lockwood, H. and Smits, S. (2011) *Supporting Rural Water Supply: Moving Towards a Service Delivery Approach*, Rugby: Practical Action Publishing.

McCommon, C., Warner, D. and Yohalem, D. (1990) *Community Management of Rural Water Supply and Sanitation Services*, Water and Sanitation Program Discussion Paper Series No. 4. Washington, DC: The World Bank. http://documents.worldbank.org/curated/en/174491468780008395/Community-management-of-rural-water-supply-and-sanitation-services (last accessed 28 January 2023).

Miller, M., Cronk, R., Klug, T., Kelly, E., Behnke, N. and Bartram, J. (2019) 'External Support Programs To Improve Rural Drinking Water Service Sustainability: A Systematic Review', *Science of Total Environment*, 670: 717–731. https://doi.org/10.1016/j.scitotenv.2019.03.069

Mugumya, F. (2013) *Enabling Community-Based Water Management Systems: Governance and Sustainability of Rural Point-water Facilities in Uganda*, PhD thesis, School of Law and Government, Dublin City University.

Mugumya, F., Ronaldo Munck, R. and Asingwire, N. (2015) 'Leveraging Community Capacity to Manage Improved Point-Water Facilities', in G.H. Fagan, S. Linnane, K.G. McGuigan and A.I. Rugumayo (eds) *Water Is Life: Progress to Secure Safe Water Provision in Rural Uganda*, Rugby: Practical Action Publishing.

Naiga, R. (2018) 'Conditions for Successful Community-based Water Management: Perspectives from Rural Uganda', *International Journal of Rural Management*, 14(2) 1–26. https://doi.org/10.1177/0973005218793245

Paul, S. (1987) *Community Participation in Development Projects: The World Bank Experience*, World Bank Discussion Paper No. 6, Washington, DC: The World Bank. http://documents.worldbank.org/curated/en/850911468766244486/Community-participation-in-development-projects-the-World-Bank-experience (last accessed 28 January 2023).

Reddy, V.R., Ramamohan Roa, M.S. and Venkataswamy, M. (2010) *'Slippage': The Bane of Rural Drinking Water Sector (A Study of Extent and Causes in Andhra Pradesh)*, WASHCost Working Paper No.06, CESS Working Paper No. 87. https://www.ircwash.org/sites/default/files/Reddy-2010-Slippage.pdf (last accessed 28 January 2023).

Schouten, T. (2006) 'Scaling Up Community Management of Rural Water Supply', WS Factsheet, WELL Resource Centre for Water, Sanitation and Environmental Health, Water Engineering and Development Centre (WEDC). https://www.lboro.ac.uk/research/wedc/well/water-supply/ws-factsheets/scaling-up-rws/ (last accessed 7 June 2024).

Schouten, T. and Moriarty, P. (2003) *Community Water, Community Management: From System to Service in Rural Areas*, Rugby: Practical Action Publishing.

UN Water (no date) 'Financing Water and Sanitation', https://www.unwater.org/water-facts/financing-water-and-sanitation (last accessed 31 December 2023).

Whaley, L. and Cleaver, F. (2017) 'Can "Functionality" Save the Community Management Model of Rural Water Supply?', *Water Resources and Rural Development*, 9: 56–66. https://doi.org/10.1016/j.wrr.2017.04.001

World Bank (2016) *Strengthening Local Providers for Improved Rural Water Supply in Pakistan*, Washington, DC: World Bank Group. http://documents.worldbank.org/curated/en/801611468332959518/Pakistan-Strengthening-local-providers-for-improved-rural-water-supply-in-Pakistan (last accessed 7 May 2024).

World Bank (2017) *Sustainability Assessment of Rural Water Service Delivery Models: Findings of a Multi-Country Review*, Washington, DC: The World Bank. https://openknowledge.worldbank.org/handle/10986/27988 (last accessed 28 January 2023).

Chapter 2 Background: Gilgit-Baltistan, WASEP, and the research

Ahmad, M., Khan, A., Tariq, S. and Blaschke, T. (2020) 'Contrasting Changes in Snow Cover and its Sensitivity to Aerosol Optical Properties in Hindukush-Karakoram-Himalaya Region', *Science of The Total Environment*, 699. https://doi.org/10.1016/j.scitotenv.2019.134356

Archer, D.R. and Fowler, H.J. (2004) 'Spatial and Temporal Variations in Precipitation in the Upper Indus Basin, Global Teleconnections and Hydrological Implications', *Hydrological and Earth Systems Sciences* 8(1): 47–61. https://doi.org/10.5194/hess-8-47-2004

British Academy (2019) 'Urban Infrastructures of Well-Being Programme: Scheme Notes for Applicants, 2019 Competition', London: The British Academy.

Grieser, A. (2018) *Den Verlauf Kontrollieren. Eine Ethnographie Zur Waterscape von Gilgit, Pakistan. Ressourcen – Gemeinschaften – Überwachung*, Bielefeld: Transcript.

Grieser, A. and Sökefeld, M. (2015) 'Intersections of Sectarian Dynamics and Spatial Mobility in Gilgit-Baltistan', in S. Conermann and E. Smolarz (eds), *Mobilizing Religion: Networks and Mobility*, Berlin: EB-Verlag, pp. 83–110.

Holden, L. (2019) 'Law and Governance in Gilgit Baltistan: Introduction', *South Asian History and Culture*, 10(1): 1–13. https://doi.org/10.1080/19472498.2019.1576300

Hunzai, I. (2013) *Special Report: Conflict Dynamics in Gilgit-Baltistan*, Washington, DC: United States Institute of Peace.

JICA (Japan International Cooperation Agency)/GoGB (Government of Gilgit-Baltistan) (2010) 'Feasibility Study (Preparatory Survey) for Gilgit Baltistan (Northern Areas) Sustainable Integrated Community Development Project', Interim Report prepared by JICA.

Sharma, E., Molden, D., Rahman, A., Khatiwada, Y.B., Zhang, L., Singh, S.P., Yao, T. and Wester, P. (2019) 'Introduction to the Hindu Kush Himalaya Assessment', in P. Wester, A. Mishra, A. Mukherji, and A. Shrestha (eds), *The Hindu Kush Himalaya Assessment*, pp. 2–16, Springer, Cham.

Chapter 3 Water governance and related institutions in Gilgit-Baltistan

Ahmed, P., Eigen-Zucchi, C., Noshab, F. and Parvez, S. (2010) *Pakistan – Gilgit-Baltistan Economic Report: Broadening the Transformation*, Report No. 74395. World Bank Group. https://documents.worldbank.org/en/publication/documents-reports/documentdetail/939151468062972670/ (last accessed 7 July 2024).

Anjum, Z.H. (2001) 'New Local Government System: A Step Towards Community Empowerment?' *The Pakistan Development Review*, 40(4): 845–867. https://pide.org.pk/research/new-local-government-system-a-step-towards-community-empowerment/ (last accessed 11 June 2024).

Broek, M. and Brown, J. (2015) 'Blueprint for Breakdown? Community Based Management of Rural Groundwater in Uganda', *Geoforum*, 67: 51–63. https://doi.org/10.1016/j.geoforum.2015.10.009

Chowns, E. (2019) 'Water Point Sustainability and the Unintended Impacts of Community Management in Malawi', in R.J. Shaw (ed.), *Water, Sanitation and Hygiene Services Beyond 2015 – Improving Access and Sustainability: Proceedings of the 38th WEDC International Conference*, Loughborough, UK, 27–31 July.

Cooper, R. (2018) *Water Management/Governance Systems in Pakistan*, K4D Helpdesk Report, Brighton: Institute of Development Studies. https://assets.publishing.service.gov.uk/media/5c6c293140f0b647b35c4393/503_Water_Governance_Systems_Pakistan.pdf (last accessed 28 January 2023).

Gilgit-Baltistan Environmental Protection Agency (GB-EPA) (2012) *Water and Wastewater Quality Survey in Gilgit Baltistan (GB)*, Gilgit: GB-EPA.

Gilgit-Baltistan Legislative Assembly (GBLA) (2014) 'The Gilgit-Baltistan Local Government Bill, 2014', Gilgit: Gilgit-Baltistan Legislative Assembly.

Government of Gilgit-Baltistan (GoGB) (2015) 'The Gilgit-Baltistan Environmental Protection Act, 2015', Gilgit: Gilgit-Baltistan Legislative Assembly.

Government of Gilgit-Baltistan (GoGB) (2019) 'Draft Drinking Water Policy', Gilgit: Gilgit-Baltistan Legislative Assembly.

Government of Pakistan (GoP) (2019a) 'Joint Sector Review – January 2019: Drinking Water, Sanitation and Hygiene, Gilgit Baltistan', Report prepared by AWF Private Limited with Government of Pakistan and UNICEF.

Government of Pakistan (GoP) (2019b) 'Joint Sector Review – July 2019: Drinking Water, Sanitation and Hygiene, Gilgit Baltistan', Report prepared by AWF Private Limited with Government of Pakistan and UNICEF.

Government of Pakistan (GoP) (2020) 'The Gilgit-Baltistan Development of Cities Act 2020', *Gazette of Pakistan*, 14 October, Islamabad.

Holden, L. (2019) 'Law and Governance in Gilgit Baltistan: Introduction', *South Asian History and Culture*, 10(1): 1–13. https://doi.org/10.1080/19472498.2019. 1576300.

Lerebours, A. and Villeminot, N. (2017) 'WASH Governance in Support of NGO Work: Trends and Differences from Field Studies' in Shaw, R.J. (ed) *Local Action with International Cooperation to Improve and Sustain Water, Sanitation and Hygiene (WASH) Services: Proceedings of the 40th WEDC International Conference, Loughborough, UK, 24–28 July 2017*, Paper 2677. https://repository.lboro.ac.uk/articles/conference_contribution/WASH_governance_in_support_of_NGO_work_trends_and_differences_from_field_studies/9589073/1 (last accessed 11 June 2024).

Muhula, R. (2019) *Pakistan @ 100: Governance & Institutions*, Policy Note March 2019, Washington, DC: The World Bank. https://documents1.worldbank.org/curated/ru/819251552645668502/pdf/Pakistan-at-100-Governance-and-Institutions.pdf (last accessed 7 May 2024).

State Bank of Pakistan (2017) *Annual Report 2016–17*, Karachi: State Bank of Pakistan. https://www.sbp.org.pk/reports/annual/arFY17/Anul-index-eng-17.htm (last accessed 23 December 2023).

UN-Water (2014) *Investing in Water and Sanitation: Increasing Access, Reducing Inequalities*, UN-Water Global Analysis and Assessment of Sanitation and Drinking-Water, GLAAS 2014 Report. https://sdgs.un.org/publications/investing-water-and-sanitation-increasing-access-reducing-inequalities-17871 (last accessed 11 June 2024).

World Bank (2016) *Strengthening Local Providers for Improved Rural Water Supply in Pakistan*, Washington, DC: World Bank Group. http://documents.worldbank.org/curated/en/801611468332959518/Pakistan-Strengthening-local-providers-for-improved-rural-water-supply-in-Pakistan (last accessed 7 May 2024).

World Bank (2017) *Sustainability Assessment of Rural Water Service Delivery Models: Findings of a Multi-Country Review*, Washington, DC: The World Bank. https://openknowledge.worldbank.org/handle/10986/27988 (last accessed 28 January 2023).

Young, W.J., Anwar, A., Bhatti, T., Borgomeo, E., Davies, S., Garthwaite III, W.R., Gilmont, E.M., Leb, C., Lytton, L., Makin, I. and Saeed, B. (2019) *Pakistan: Getting More from Water*, Washington DC: The World Bank. https://hdl.handle.net/10986/31160 (last accessed 7 May 2024).

Chapter 4 The WASEP model of community management

Aga Khan Health Service (AKHS) (1997) 'Water, Sanitation, Hygiene, and Health Study Project (WSHHSP)'. Report prepared by AKHS Northern Areas and Chitral.

Aga Khan Planning and Building Service, Pakistan (AKPBS-P) (n.d.1) 'WASEP: Water and Sanitation Extension Programme, Programme Cycle 1997 to 2001'. AKPBS: Karachi. https://www.ircwash.org/resources/wasep-water-and-sanitation-extension-programme-project-aga-khan-planning-and-building (last accessed 11 June 2024).

Aga Khan Planning and Building Service, Pakistan (AKPBS-P) (n.d.2) 'Immediate Assessment Report: Community Physical Infrastructure Projects', Karachi: AKPBS-P.

Aga Khan Planning and Building Service, Pakistan (AKPBS-P) (2012) 'Standard Operating Procedures: Water and Sanitation Extension Program'. Report by AKPBS-P, June 2012.

Aga Khan Rural Support Programme (AKRSP) (2011) 'LSO Directory (A Directory of the Federations of the Village/Women Organizations in Gilgit-Baltistan and Chitral)', April. Compiled and edited by Beg, G.A., Khan, Z.A. and Aftab, M. Funded by Canadian International Development Agency (CIDA) under Institutional Development for Poverty Alleviation (IDPR).

Ahmed, J. and Alibhai, K. (2000) 'Community Management RWSS in Northern Pakistan' in Pickford, J. (ed) *Water, Sanitation and Hygiene – Challenges of the Millennium: Proceedings of the 26th Water Engineering Development Centre (WEDC) Conference, Dhaka, Bangladesh, 5–9 November*, pp. 61–65. https://repository.lboro.ac.uk/articles/conference_contribution/

Community_management_of_RWSS_in_Northern_Pakistan/9593537/1 (last accessed 11 June 2024).

Ahmed, J., Langendijk, M., Murad, F. and Hussain, M. (1996) 'A Water and Sanitation Inventory of 862 villages of Northern Areas and Chitral', Issue Paper 8: Water, Sanitation, Hygiene, and Health Study Project (WSHHSP) October. Gilgit: Aga Khan Health Service Pakistan.

Clemens, J. (2000) 'Rural Development in Northern Pakistan: Impacts of the Aga Khan Rural Support Programme', in A. Dittmann (ed.), *Mountain Societies in Transition: Contributions to the Cultural Geography of the Karakorum*, Köln: Rüdiger Köppe Verlag, pp. 1–35.

Grieser, A. (2018) *Den Verlauf Kontrollieren. Eine Ethnographie Zur Waterscape von Gilgit, Pakistan. Ressourcen – Gemeinschaften – Überwachung*, Bielefeld: Transcript.

Hussain, A. and Langendijk, M. (1995) 'Self-Help Rural Water Supply Schemes: Lessons Learned from the Northern Areas of Pakistan', Issue Paper 4: Water, Sanitation, Hygiene and Health Studies Project (WSHHSP). Gilgit: Aga Khan Health Service Pakistan.

Hussain, M., Khan, S. and Alibhai, K. (2000) 'Water Tariffs: A Challenging Issue for WASEP Implementation', in Pickford, J. (ed) *Water, Sanitation and Hygiene – Challenges of the Millennium: Proceedings of the 26th Water Engineering Development Centre (WEDC) Conference, Dhaka, Bangladesh, 5–9 November*, pp. 76–80. https://repository.lboro.ac.uk/articles/conference_contribution/Water_tariffs_a_challenging_issue_for_WASEP_implementation/9593849 (last accessed 11 June 2024).

Khan, H.W. and Hunzai, I. (2000) 'Bridging Institutional Gaps in Irrigation Management: The Post "Ibex-Horn" Innovations in Northern Pakistan', in H. Kreutzmann (ed.), *Sharing Water: Irrigation and Water Management in the Hindukush-Karakoram-Himalaya*, Oxford: Oxford University Press, pp. 133–145.

Korput, J.A., Muneeba, and Langendijk, M. (1995) 'Hygiene Behaviour in North Pakistan: The Results of a Quantitative Study', Issue Paper 6: Water, Sanitation, Hygiene, and Health Study Project (WSHHSP), December. https://www.ircwash.org/resources/hygiene-behaviour-north-pakistan-results-quantitative-and-qualitative-study (last accessed 11 June 2024).

Langendijk, M., Korput, J.A., Raza, H. and Ahmed, K. (1996) 'A Study on Behavioural and Microbiological Aspects of Handwashing in Northern Pakistan: The Development of Appropriate Handwashing Messages', Issue Paper 9: Water, Sanitation, Hygiene, and Health Study Project (WSHHSP), October. https://www.ircwash.org/resources/study-behavioural-and-microbiological-aspects-handwashing-northern-pakistan-development (last accessed 11 June 2024).

Miller, K.J.L. (2015) *A Spiritual Development: Islam, Volunteerism and International Development in the Hunza Valley, Northern Pakistan*, PhD dissertation, University of California, San Diego.

Sökefeld, M. (1997) 'Ein Labyrinth von Identitäten in Nordpakistan', *Culture Area Karakorum Scientific Studies*, Köln: Rüdiger Köppe Verlag.

Chapter 5 Community participation in WASEP

Aga Khan Planning and Building Service, Pakistan (AKPBS-P) (2012) 'Standard Operating Procedures: Water and Sanitation Extension Program'. Report by AKPBS-P, June 2012.

Birkinshaw, M., Grieser, A. and Tan, J. (2021) 'How Does Community-Managed Infrastructure Scale Up from Rural to Urban? An Example of Co-Production in Community Water Projects in Northern Pakistan', *Environment & Urbanization*, 33(2): 496–518. https://doi.org/10.1177/09562478211034853

Cleaver, F. (1999) 'Paradoxes of Participation: Questioning Participatory Approaches to Development', *Journal of International Development: The Journal of the Development Studies Association*, 11(4): 597–612. https://doi.org/10.1002/(SICI)1099-1328(199906)11:4<597::AID-JID610>3.0.CO;2-Q

Grieser, A. (2018) *Den Verlauf Kontrollieren. Eine Ethnographie Zur Waterscape von Gilgit, Pakistan. Ressourcen – Gemeinschaften – Überwachung*, Bielefeld: Transcript.

Hutchings, P., Chan, M.Y., Cuadrado, L., Ezbakhe, F., Mesa, B., Tamekawa, C. and Franceys, R. (2015) 'A Systematic Review of Success Factors in the Community Management of Rural Water Supplies Over the Past 30 Years', *Water Policy*, 17(5): 963–983. https://doi.org/10.2166/wp.2015.128

Hutchings, P., Franceys, R., Mekala, S., Smits, S. and James, A.J. (2016) 'Revisiting the History, Concepts and Typologies of Community Management for Rural Drinking Water Supply in India', *International Journal of Water Resources Development* 33(1): 152–169. https://doi.org/10.1080/07900627.2016.1145576

McCommon, C., Warner, D. and Yohalem, D. (1990) *Community Management of Rural Water Supply and Sanitation Services*, Water and Sanitation Program Discussion Paper Series No. 4. Washington, DC: The World Bank. http://documents.worldbank.org/curated/en/174491468780008395/Community-management-of-rural-water-supply-and-sanitation-services (last accessed 28 January 2023).

Schouten, T. and Moriarty, P. (2003) *Community Water, Community Management: From System to Service in Rural Areas*, Rugby: Practical Action Publishing.

Whaley, L. and Cleaver, F. (2017) 'Can "Functionality" Save the Community Management Model of Rural Water Supply?', *Water Resources and Rural Development*, 9: 56–66. https://doi.org/10.1016/j.wrr.2017.04.001

World Bank (2017) *Sustainability Assessment of Rural Water Service Delivery Models: Findings of a Multi-Country Review*, Washington, DC: The World Bank. https://openknowledge.worldbank.org/handle/10986/27988 (last accessed 28 January 2023).

Chapter 6 Women's participation in WASEP

Aga Khan Foundation (AKF) and Aga Khan Planning and Building Service, Pakistan (AKPBS-P) (2014) 'KfW Development Bank Funded Water and Sanitation Extension Programme in Chitral and Gilgit-Baltistan, Closeout Report, August 2010–June 2014'. Report prepared by AKF and AKPBS-P.

Aga Khan Planning and Building Service, Pakistan (AKPBS-P) and USAID (2016) 'Water and Hygiene Improvement Project (WHIP), Murtazaabad, Hunza, 1st Annual Report, September, 2015–August, 2016'. Report prepared by AKPBS-P for USAID, Agreement No. AID-391-A-15-00007.

Ahmed, J. and Alibhai, K. (2000) 'Community Management RWSS in Northern Pakistan', in Pickford, J. (ed) *Water, Sanitation and Hygiene – Challenges of the Millennium: Proceedings of the 26th Water Engineering Development Centre (WEDC) Conference, Dhaka, Bangladesh, 5–9 November*, pp. 61–65.

https://repository.lboro.ac.uk/articles/conference_contribution/Community_management_of_RWSS_in_Northern_Pakistan/9593537/1 (last accessed 11 June 2024).

Alibhai, K., Ahmad, T. and Aziz, N. (2001) 'Evolution of Women's Involvement in Projects in Northern Pakistan', in Scott, R. (ed). *People and Systems for Water, Sanitation and Health: Proceedings of the 27th Water Engineering Development Centre (WEDC) Conference, Lusaka, Zambia, 20–24 August*, pp. 513–516. https://repository.lboro.ac.uk/articles/conference_contribution/Evolution_of_women_s_involvement_in_projects_in_N_Pakistan/9591887 (last accessed 11 June 2024).

Barnett, T., Ahmad, T., Baig, Y. and Alibai, K. (2001) 'WASEP's Role in Improving Women's Participation in WSS Projects', *Proceedings of the 27th Water Engineering Development Centre (WEDC) Conference, Lusaka, Zambia, 20–24 August.* http://wedc.lboro.ac.uk/resources/conference/27/Barnett.pdf (last accessed 14 June 2024).

Besio, K. (2007) 'Depth of Fields: Travel Photography and Spatializing Modernities in Northern Pakistan', *Environment and Planning D: Society and Space*, 25(1): 53–74. https://doi.org/10.1068/d2504

Birkinshaw, M., Grieser, A. and Tan, J. (2021) 'How Does Community-Managed Infrastructure Scale Up from Rural to Urban? An Example of Co-Production in Community Water Projects in Northern Pakistan', *Environment & Urbanization*, 33(2): 496–518. https://doi.org/10.1177/09562478211034853

Carrard, N., Crawford, J., Halcrow, G., Rowland, C. and Willetts, J. (2013) 'A Framework for Exploring Gender Equality Outcomes from WASH Programmes', *Waterlines*, 32(4): 315–333. https://doi.org/10.3362/1756-3488.2013.033

Cleaver, F. (1991) 'Maintenance of Rural Water Supplies in Zimbabwe: User Participation in Zimbabwe through Waterpoint Committees Promotes Handpump Sustainability', *Waterlines*, 9(4): 23–26. https://doi.org/10.3362/0262-8104.1991.017

Cleaver, F. (1998) 'Incentives and Informal Institutions: Gender and the Management of Water', *Agriculture and Human Values*, 15: 347–36. https://doi.org/10.1023/A:1007585002325

Cleaver, F. (2000) 'Analysing Gender Roles in Community Natural Resource Management: Negotiation, Lifecourses and Social Inclusion', *IDS Bulletin*, 31(2): 1–11. https://doi.org/10.1111/j.1759-5436.2000.mp31002008.x

Fisher, J. (2006) *For Her It's the Big Issue: Putting Women at the Centre of Water Supply, Sanitation and Hygiene*, Water, Sanitation and Hygiene Evidence Report, Geneva: Water Supply and Sanitation Collaborative Council.

Gratz, K. (2006) *Verwandtschaft, Geschlecht und Raum: Aspekte weiblicher Lebenswelt in Gilgit/Nordpakistan.* Köln: Köppe.

Grieser, A. (2018) *Den Verlauf Kontrollieren. Eine Ethnographie Zur Waterscape von Gilgit, Pakistan. Ressourcen – Gemeinschaften – Überwachung*, Bielefeld: Transcript.

Halvorson, S.J., Aziz, N. and Alibai, K. (1998) 'Strategies to Involve Women in Water Supply and Sanitation' in Pickford, J. (ed). *Sanitation and Water for All: Proceedings of the 24th WEDC International Conference, Islamabad, Pakistan, 31 August–4 September*, pp. 233–236. https://repository.lboro.ac.uk/articles/conference_contribution/Strategies_to_involve_women_in_water_supply_and_sanitation/9597200 (last accessed 14 June 2024).

Hannah, C., Giroux, S., Krell, N., Lopus, S., McCann, L.E., Zimmer, A., Caylor, K.K. and Evans, T.P. (2021) 'Has the Vision of a Gender Quota Rule Been Realized for Community-Based Water Management Committees in Kenya?' *World Development*, 137: 1–13. https://doi.org/10.1016/j.worlddev.2020.105154

Hemson, D. (2002) 'Women Are Weak When They Are Amongst Men: Women's Participation in Rural Water Committees in South Africa', *Agenda: Empowering Women for Gender Equity*, 52: 24–32. https://doi.org/10.2307/4066469

Hewitt, F. (1989) 'Woman's Work, Woman's Place: The Gendered Life-World of a High Mountain Community in Northern Pakistan', *Mountain Research and Development*, 9(4): 335–352. https://doi.org/10.2307/3673583

Hussain, A. and Langendijk, M. (1995) 'Self-Help Rural Water Supply Schemes: Lessons Learned from the Northern Areas of Pakistan', Issue Paper 4: Water, Sanitation, Hygiene and Health Studies Project (WSHHSP). Gilgit: Aga Khan Health Service Pakistan.

Hussain, M., Khan, S. and Alibhai, K. (2000) 'Water Tariffs: A Challenging Issue for WASEP Implementation', in Pickford, J. (ed.) *Water, Sanitation and Hygiene – Challenges of the Millennium: Proceedings of the 26th Water Engineering Development Centre (WEDC) Conference, Dhaka, Bangladesh, 5–9 November*, pp. 76–80. https://repository.lboro.ac.uk/articles/conference_contribution/Water_tariffs_a_challenging_issue_for_WASEP_implementation/9593849 (last accessed 11 June 2024).

Khandker, V., Gandhi, V.P. and Johnson, N. (2020) 'Gender Perspective in Water Management: The Involvement of Women in Participatory Water Institutions of Eastern India', *Water* 12(1): 196. https://doi.org/10.3390/w12010196

Kreutzmann, H. (1988) 'Oases of the Karakorum: Evolution of Irrigation and Social Organization in Hunza, North Pakistan' in: Allan, N.J.R. (ed.) *Human Impact on Mountains*, Totowa, New Jersey: Rowman and Littlefield.

Mandara, C.G., Niehof, A. and Horst, H. (2017) 'Women and Rural Water Management: Token Representatives or Paving the Way to Power?' *Water Alternatives* 10(1): 116–133.

Meinzen-Dick, R. and Zwarteveen, M. (1998) 'Gendered Participation in Water Management: Issues and Illustrations from Water Users' Associations in South Asia', *Agriculture and Human Values*, 15: 337–345. https://doi.org/10.1023/A:1007533018254

Mommen, B., Humphries-Waa, K. and Gwavuya, S. (2017) 'Does Women's Participation in Water Committees Affect Management and Water System Performance in Rural Vanuatu?' *Waterlines*, 36(3): 216–232. https://doi.org/0.3362/1756–3488.16–00026

Oakley, P. (1991) *Projects with People: The Practice of Participation in Rural Development*, Geneva: International Labour Organization.

Prokopy, L.S. (2004) 'Women's participation in Rural Water Supply Projects in India: Is it Moving Beyond Tokenism and Does it Matter?' *Water Policy*, 6: 103–116. https://doi.org/10.2166/wp.2004.0007

Ray, I. (2007) 'Women, Water, and Development', *Annual Review of Environment and Resources*, 32: 421–449. https://doi.org/10.1146/annurev.energy.32.041806.143704

Schnegg, M. and Linke, T. (2016) 'Travelling Models of Participation: Global Ideas and Local Translations of Water Management in Namibia', *International Journal of the Commons*, 10(2): 800–820. https://doi.org/10.18352/ijc.705

Schouten, T. and Moriarty, P. (2003) *Community Water, Community Management: From System to Service in Rural Areas*, Rugby: Practical Action Publishing.
Singh, N. (2008) 'Equitable Gender Participation in Local Water Governance: An Insight Into Institutional Paradoxes', *Water Resources Management*, 22: 925–942. https://doi.org/10.1007/s11269–007–9202-z
United Nations Department of Economic and Social Affairs (UN-DESA) (2005) *Women and Water*, New York: United Nations. https://www.un.org/womenwatch/daw/public/Feb05.pdf (last accessed 06 August 2021).
Uphoff, N., Esman, M.J. and Krishna, A. (1998) *Reasons for Success: Learning from Instructive Experiences in Rural Development*, New Delhi: Vistaar Publications.
Van Wijk-Sijbesma, C. (1997) 'Drinking Water and Sanitation: Women Can Do Much', *World Health Forum*, 8: 28–33.
Van Wijk-Sijbesma, C. (1998) *Gender in Water Resources Management, Water Supply and Sanitation: Roles and Realities Revisited*, Technical Paper Series 33-E IRC, The Hague: IRC International Water and Sanitation Centre.
Van Wijk-Sijbesma, C. (2001) *The Best of Two Worlds? Methodology for Participatory Assessment of Community Water Services*, PhD thesis, Wageningen Universiteit. https://library.wur.nl/WebQuery/wurpubs/fulltext/139858 (last accessed 6 August 2021).
Were, E., Roy, J. and Swallow, B. (2008) 'Local Organisation and Gender in Water Management: A Case Study from the Kenya Highlands', *Journal of International Development*, 20: 69–81. https://doi.org/10.1002/jid.1428
Whaley, L. and Cleaver, F. (2017) 'Can "Functionality" Save the Community Management Model of Rural Water Supply?', *Water Resources and Rural Development*, 9: 56–66. https://doi.org/10.1016/j.wrr.2017.04.001
Zwarteveen, M. (2008) 'Men, Masculinities and Water Powers in Irrigation', *Water Alternatives* 1(1): 111–130.
Zwarteveen, M. and Neupane, N. (1996) *Free-Riders or Victims: Women's Nonparticipation in Irrigation Management in Nepal's Chhattis Mauja Irrigation Scheme*, Research Report 7. Colombo, Sri Lanka: International Irrigation Management Institute.

Chapter 7 Water-related conflict and conflict management in WASEP

Aga Khan Foundation (AKF) and Aga Khan Planning and Building Service, Pakistan (AKPBS-P) (2014) 'KfW Development Bank Funded Water and Sanitation Extension Programme in Chitral and Gilgit-Baltistan, Closeout Report, August 2010–June 2014. Report prepared by AKF and AKPBS-P.
Aga Khan Rural Support Programme (AKRSP) (2011) 'LSO Directory (A Directory of the Federations of the Village/Women Organizations in Gilgit-Baltistan and Chitral)', April. Compiled and edited by Beg, G.A., Khan, Z.A. and Aftab, M. Funded by Canadian International Development Agency (CIDA) under Institutional Development for Poverty Alleviation (IDPR).
Beg, F.A. (2018) 'Social and Political Governance and Development: Perceptions and Experiences of Ibrahim Khan from Gilgit region in the Northern Pakistan' [blog], 19 October 2018. https://fazalamin.com/social-and-political-governance-and-development-perceptions-and-experiences-of-ibrahim-khan-from-gilgit-region-in-the-northern-pakistan/ (last accessed 10 July 2020).

Boelens, R. (1998) 'Equity and Rule-Making', in R. Boelens and G. Dávila (eds), *Searching for Equity: Conceptions of Justice and Equity in Peasant Irrigation*, Assen: Van Gorcum, pp. 16–36.

Boelens, R. (2009) 'The Politics of Disciplining Water Rights', *Development and Change*, 40(2): 307–331. https://doi.org/10.1111/j.1467-7660.2009.01516.x

Broek, M. (2017) *A Critical Evaluative Enquiry of the Community Based Management Model and Alternative Approaches for Sustainable Rural Water Management*, PhD thesis, University of Portsmouth, UK.

Datoo, F. (2012) 'Preliminary Technical Audit Report', Prepared for Aga Khan Planning and Building Service, Pakistan (AKPBS-P) for the Water and Sanitation Extension Program (WASEP), February–March.

Grieser, A. (2018) *Den Verlauf Kontrollieren. Eine Ethnographie Zur Waterscape von Gilgit, Pakistan. Ressourcen – Gemeinschaften – Überwachung*, Bielefeld: Transcript.

Hill, J. (2012) *A Post-Area Studies Approach to the Study of Hill Irrigation Across the Alai – Pamir – Karakoram – Himalaya*, Crossroads Asia Working Paper Series, No. 3. Bonn: Competence Network Crossroads Asia: Conflict – Migration – Development.

Hussain, A. and Langendijk, M. (1995) 'Self-Help Rural Water Supply Schemes: Lessons Learned from the Northern Areas of Pakistan', Issue Paper 4: Water, Sanitation, Hygiene and Health Studies Project (WSHHSP). Gilgit: Aga Khan Health Service Pakistan.

Kreutzmann, H. (2000a) 'Water Management in Mountain Oases of the Karakoram', in H. Kreutzmann (ed.), *Sharing Water: Irrigation and Water Management in the Hindukush-Karakoram-Himalaya*, Oxford: Oxford University Press, pp. 91–115.

Kreutzmann, H. (2000b) 'Water Towers of Humankind: Approaches and Perspectives for Research on Hydraulic Resources in the Mountains of South and Central Asia', in H. Kreutzmann (ed.), *Sharing Water: Irrigation and Water Management in the Hindukush-Karakoram-Himalaya*, Oxford: Oxford University Press, pp. 13–31.

Miller, K.J.L. (2015) *A Spiritual Development: Islam, Volunteerism and International Development in the Hunza Valley, Northern Pakistan*, PhD dissertation, University of California, San Diego.

Narain, V. (2008) 'Warabandi as a Sociotechnical System for Canal Water Allocation: Opportunities and Challenges for Reform', *Water Policy*, 10: 409–422. https://doi.org/10.2166/wp.2008.057

Sökefeld, M. (1997) 'Ein Labyrinth von Identitäten in Nordpakistan', *Culture Area Karakorum Scientific Studies*, Köln: Rüdiger Köppe Verlag.

Sökefeld, M. (1998) 'The People Who Really Belong to Gilgit: Theoretical and Ethnographical Perspectives on Identity and Conflict', in I. Stellrecht and H.G. Bohle (eds), *Transformation of Social and Economic Relationships in Northern Pakistan*, Köln: Rüdiger Köppe Verlag, pp. 97–222.

Trawick, P. (2001) 'The Moral Economy of Water: Equity and Antiquity in the Andean Commons', *American Anthropologist*, 103(2): 361–379.

Wutich, A., Brewis, A., Sigurdsson, S., Stotts, R. and York, A. (2015) 'Fairness and the Human Right to Water: A Preliminary Cross-Cultural Theory', in J.R. Wagner (ed.), *The Social Life of Water*, New York: Berghahn, pp. 220–238.

Chapter 8 How financially sustainable is WASEP?

Aga Khan Planning and Building Service, Pakistan (AKPBS-P) (no date) 'WASEP: Water and Sanitation Extension Programme, Programme Cycle 1997 to 2001'. Karachi: AKPBS-P. https://www.ircwash.org/resources/wasep-water-and-sanitation-extension-programme-project-aga-khan-planning-and-building (last accessed 11 June 2024).

Aga Khan Planning and Building Service, Pakistan (AKPBS-P) and USAID (2016) 'Water and Hygiene Improvement Project (WHIP), Murtazaabad, Hunza, 1st Annual Report, September, 2015–August, 2016'. Report prepared by AKPBS-P for USAID, Agreement No. AID-391–A-15–00007.

Fonseca, C., Franceys, R., Batchelor, C., McIntyre, P., Klutse, A., Komives, K., Moriarty, P., Naafs, A., Nyarko, K., Pezon, C., Potter, A., Reddy, R. and Snehalatha, M. (2011) *Life-Cycle Costs Approach: Costing Sustainable Services*, WASHCost Briefing Note 1a, The Hague: IRC International Water and Sanitation Centre. https://www.ircwash.org/resources/briefing-note-1a-life-cycle-costs-approach-costing-sustainable-service (last accessed 28 January 2023).

Hope, R., Thomson, P., Koehler, J. and Foster, T. (2020) 'Rethinking the Economics of Rural Water in Africa', *Oxford Review of Economic Policy*, 36(1): 171–190. https://doi.org/10.1093/oxrep/grz036

Hussain, M., Khan, S. and Alibhai, K. (2000) 'Water Tariffs: A Challenging Issue for WASEP Implementation', in Pickford, J. (ed.) *Water, Sanitation and Hygiene – Challenges of the Millennium: Proceedings of the 26th Water Engineering Development Centre (WEDC) Conference, Dhaka, Bangladesh, 5–9 November*, pp. 76–80. https://repository.lboro.ac.uk/articles/conference_contribution/Water_tariffs_a_challenging_issue_for_WASEP_implementation/9593849 (last accessed 11 June 2024).

Lockwood, H. and Smits, S. (2011) *Supporting Rural Water Supply: Moving Towards a Service Delivery Approach*, Rugby: Practical Action Publishing.

Majuru, B., Suhrcke, M. and Hunter, P.R. (2018) 'Reliability of Water Supplies in Low and Middle-Income Countries: A Structured Review of Definitions and Assessment Criteria', *Journal of Water, Sanitation and Hygiene for Development*, 8(2): 142–164. https://doi.org/10.2166/washdev.2018.174

Mugumya, F. (2013) *Enabling Community-Based Water Management Systems: Governance and Sustainability of Rural Point-water Facilities in Uganda*, PhD thesis, School of Law and Government, Dublin City University.

Schouten, T. and Moriarty, P. (2003) *Community Water, Community Management: From System to Service in Rural Areas*, Rugby: Practical Action Publishing.

Torbaghan, M.E. and Burrow, M. (2019) *Evidence of Small Town Urban Water Supply Costs per Capita and Unit Costs*, K4D Helpdesk Report. Brighton, UK: Institute of Development Studies.

World Bank (2017) *Sustainability Assessment of Rural Water Service Delivery Models: Findings of a Multi-Country Review*, Washington, DC: The World Bank. https://openknowledge.worldbank.org/handle/10986/27988 (last accessed 28 January 2023).

World Health Organization and UN Water (2014) *UN-Water Global Analysis and Assessment of Sanitation and Drinking-Water (GLAAS) 2014 Report: Investing in Water and Sanitation: Increasing Access, Reducing Inequalities*. Geneva: World Health Organization.

Chapter 9 WASEP water infrastructure and water quality

Aga Khan Planning and Building Service, Pakistan (AKPBS-P) (no date) 'WASEP: Water and Sanitation Extension Programme, Programme Cycle 1997 to 2001'. Karachi: AKPBS. https://www.ircwash.org/resources/wasep-water-and-sanitation-extension-programme-project-aga-khan-planning-and-building (last accessed 11 June 2024).

Aga Khan Planning and Building Service, Pakistan (AKPBS-P) (2012) 'Standard Operating Procedures: Water and Sanitation Extension Program'. Report by AKPBS-P, June.

Ahsan, M., Rasheed, H., Ashraf, M. and Anwaar, K. (2021) 'Assessment of Water Quality Status in Gilgit-Baltistan', Pakistan Council of Research in Water Resources (PCRWR). Islamabad: PCRWR.

Gilgit-Baltistan Environmental Protection Agency (GB-EPA) (2012) *Water and Wastewater Quality Survey in Gilgit Baltistan (GB)*, Gilgit: GB-EPA.

Gilgit-Baltistan Environmental Protection Agency (GB-EPA) (2013) *Water and Wastewater Quality Survey in Seven Urban Centres of Gilgit Baltistan*, Gilgit: GB-EPA.

Gilgit-Baltistan Environmental Protection Agency (GB-EPA) (2019) *Assessment of Drinking Water Quality 'Natural Springs and Surface Water' in Gilgit-Baltistan–2019*, Gilgit: GB-EPA.

Government of Pakistan (GoP) (2019) 'Joint Sector Review – July 2019: Drinking Water, Sanitation and Hygiene, Gilgit Baltistan', Report prepared by AWF Private Limited with Government of Pakistan and UNICEF.

Hussain, S.W., Hussain, K., Zehra, Q., Liaqat, S., Ali, A., Abbas, Y. and Hussain, B. (2022) 'Assessment of Drinking Water Quality "Natural Springs and Surface Water" and Associated Health Risks in Gilgit-Baltistan Pakistan', *Pure and Applied Biology*, 11(4): 919–931. https://doi.org/10.19045/bspab. 2022.110095

Schouten, T. and Moriarty, P. (2003) *Community Water, Community Management: From System to Service in Rural Areas*, Rugby: Practical Action Publishing.

Chapter 10 Natural hazards: WASEP engineering solutions and community responses

Aga Khan Planning and Building Service, Pakistan (AKPBS-P) (2012) 'Standard Operating Procedures: Water and Sanitation Extension Program', Report by AKPBS-P, June 2012.

Baig, S.U., Rehman, M.U. and Janjua, N.N. (2021) 'District-Level Disaster Risk and Vulnerability in the Northern Mountains of Pakistan', *Geomatics, Natural Hazards and Risk*, 12(1): 2002–2022. https://doi.org/10.1080/19475705.2021. 1944331

Hussain, S.W., Hussain, K., Zehra, Q., Liaqat, S., Ali, A., Abbas, Y. and Hussain, B. (2022) 'Assessment of Drinking Water Quality "Natural Springs and Surface Water" and Associated Health Risks in Gilgit-Baltistan Pakistan', *Pure and Applied Biology*, 11(4): 919–931. https://doi.org/10.19045/bspab.2022.110095

Shah, A., Ali, K., Nizami, S.M., Jan, I.U., Hussain, I., Begum, F. and Khan, H. (2019) 'Risk Assessment of Shishper Glacier, Hassanabad Hunza, North Pakistan', *Journal of Himalayan Earth Sciences* 52(1): 1–11.

Chapter 11 How sustainable and scalable is WASEP?

Broek, M. (2009) *A Critical Evaluative Enquiry of the Community Based Management Model and Alternative Approaches for Sustainable Rural Water Management*, PhD thesis, University of Portsmouth.

Broek, M. and Brown, J. (2015) 'Blueprint for Breakdown? Community Based Management of Rural Groundwater in Uganda', *Geoforum*, 67: 51–63. https://doi.org/10.1016/j.geoforum.2015.10.009

Carter, R. (2021) *Rural Community Water Supply: Sustainable Services for All*, Rugby: Practical Action Publishing.

Chowns, E. (2019) 'Water Point Sustainability and the Unintended Impacts of Community Management in Malawi', in R.J. Shaw (ed.), *Water, Sanitation and Hygiene Services Beyond 2015 – Improving Access and Sustainability: Proceedings of the 38th WEDC International Conference*, Loughborough, UK, 27–31 July.

Fonseca, C. (2015) 'Financing Universal Access: The Role of Water Financing Facilities' [blog], 21 June 2015, International Water and Sanitation Centre (IRC). https://www.ircwash.org/blog/financing-universal-access-role-water-financing-facilities (last accessed 17 March 2024).

Hannah, C., Giroux, S., Krell, N., Lopus, S., McCann, L.E., Zimmer, A., Caylor, K.K. and Evans, T.P. (2021) 'Has the Vision of a Gender Quota Rule Been Realized for Community-Based Water Management Committees in Kenya?' *World Development*, 137: 1–13. https://doi.org/10.1016/j.worlddev.2020.105154

Hutchings, P., Chan, M.Y., Cuadrado, L., Ezbakhe, F., Mesa, B., Tamekawa, C. and Franceys, R. (2015) 'A Systematic Review of Success Factors in the Community Management of Rural Water Supplies over the Past 30 Years', *Water Policy*, 17(5): 963–983. https://doi.org/10.2166/wp.2015.128

Hutchings, P., Franceys, R., Mekala, S., Smits, S. and James, A.J. (2016) 'Revisiting the History, Concepts and Typologies of Community Management for Rural Drinking Water Supply in India', *International Journal of Water Resources Development* 33(1): 152–169. https://doi.org/10.1080/07900627.2016.1145576

Lockwood, H. and Smits, S. (2011) *Supporting Rural Water Supply: Moving Towards a Service Delivery Approach*, Rugby: Practical Action Publishing.

McCommon, C., Warner, D. and Yohalem, D. (1990) *Community Management of Rural Water Supply and Sanitation Services*, Water and Sanitation Program Discussion Paper Series No. 4. Washington, DC: The World Bank. http://documents.worldbank.org/curated/en/174491468780008395/Community-management-of-rural-water-supply-and-sanitation-services (last accessed 28 January 2023).

Schouten, T. and Moriarty, P. (2003) *Community Water, Community Management: From System to Service in Rural Areas*, Rugby: Practical Action Publishing.

Soto, A., Macura, B., Del Duca, L., Carrard, N., Gosling, L., Hannes, K., Thomas, J., Sara, L., Sommer, M., Waddington, H.S., Foggitt, E., Njoroge, G., Fadhila, A., Orlando, A. and Dickin, S. (2021) 'Do Rural Water Supply Interventions Contribute to Gender and Social Equality? A Mixed Methods Systematic Review', in *42nd Water Engineering Development Centre (WEDC) International Conference* [online], 13–15 September.

Tan, J. (2008) *Privatization in Malaysia: Regulation, Rent-Seeking and Policy Failure*, London: Routledge.

Tan, J. (2011) 'Infrastructure Privatisation: Oversold, Misunderstood and Inappropriate', *Development Policy Review*, 29(1): 47–74. https://doi.org/10.1111/j.1467-7679.2011.00513.x

UN Water (no date) 'Financing Water and Sanitation', https://www.unwater.org/water-facts/financing-water-and-sanitation (last accessed 31 December 2023).

World Bank (2017) *Sustainability Assessment of Rural Water Service Delivery Models: Findings of a Multi-Country Review*, Washington, DC: The World Bank. https://openknowledge.worldbank.org/handle/10986/27988 (last accessed 28 January 2023).

Index

C

E

N

O

P

R

S

www.ingramcontent.com/pod-product-compliance
Lightning Source LLC
Jackson TN
JSHW070235140825
89344JS00021B/571